RELATIVITY AND
HIGH ENERGY PHYSICS

THE WYKEHAM SCIENCE SERIES
for schools and universities

General Editors:

Professor Sir Nevill Mott, F.R.S.
Cavendish Professor of Physics
University of Cambridge

G. R. Noakes
Formerly Senior Physics Master
Uppingham School

To broaden the outlook of the senior grammar school pupil and to introduce the undergraduate to the present state of science as a university study is the aim of the Wykeham Science Series. Each book seeks to reinforce this link between school and university levels, and the main author, a university teacher distinguished in the field, is assisted by an experienced sixth-form schoolmaster.

RELATIVITY AND
HIGH ENERGY PHYSICS

W. G. V. Rosser – University of Exeter

Schoolmaster author:
R. K. McCulloch – Chorlton High School

DISTRIBUTED BY

 Springer-Verlag New York Inc.
175 Fifth Avenue, New York, N. Y. 10010

 WYKEHAM PUBLICATIONS (LONDON) LTD
(A subsidiary of Taylor & Francis Ltd)
LONDON & WINCHESTER
1969

First published 1969 by Wykeham Publications (London) Ltd.

Cover illustration – 7 GeV Proton Synchroton NIMROD – by courtesy of the Rutherford High Energy Laboratory, Chilton, Didcot, Berkshire

*Printed in Great Britain by Taylor & Francis Ltd.
10–14 Macklin Street, London, W.C.2*

85109 080 X

World wide distribution, excluding the Western Hemisphere, India, Pakistan and Japan, by Associated Book Publishers Ltd., London and Andover.

PREFACE

THE main object of the book is to develop the theory of special relativity in a way suitable for students taking A level Physics. Some more searching sections are included for the benefit of the more advanced reader. These sections are marked with an asterisk, and can be omitted in a first reading. In Chapter 2, the variation of mass with velocity is developed from experiments, before the Lorentz transformations are introduced. This leads to the development from experiments of the laws of high speed particles, and the equivalence of mass and energy. It is then shown that, if these experimental laws are to hold in all inertial reference frames, the Galilean transformations must be replaced by the Lorentz transformations. Since the experimental deviations from Newtonian mechanics are very large at speeds comparable to the speed of light, students are quite ready to accept that, at very high speeds, Newtonian mechanics, and concepts based on Newtonian mechanics, such as the Galilean transformations, must be changed in favour of the theory of special relativity.

A brief review of the historical development of the theory of special relativity is given in Chapter 3. In Chapter 4, the theory of special relativity is developed using radar methods and the K-calculus, a method pioneered by Professor H. Bondi. Some applications of the Lorentz transformations are given in Chapters 4 and 5. Accounts of rotating reference frames and the principle of equivalence are given in Chapter 6. An account of the Clock Paradox is given in Chapter 7. A few problems are given at the end of most chapters, to give the reader practice in the use of the equations of special relativity. S.I. units are used throughout. In the text the units are generally written out in full.

I would like to thank Mr. R. K. McCulloch for his invaluable advice and cooperation at every stage in the development of the book.

Exeter,   W. G. V. ROSSER
July, 1969.

UNITS AND SYMBOLS

Basic SI units

Physical quantity	Name of unit	Symbol for unit
length	metre	m
mass	kilogramme	kg
time	second	s
electric current	ampere	A

The *metre* (m) is equal to 1 650 763·73 wavelengths of the orange red line of the krypton-86 atom.

The *kilogramme* (kg) is defined as the mass of a cylinder of platinum–iridium kept at Sèvres, near Paris.

The *second* (s) is the duration of 9 192 631 770 periods of the radiation corresponding to the transition between the two hyperfine levels of the ground state of the caesium-133 atom.

The *ampere* (A) is defined as that unvarying current which, if present in each of two infinitely thin parallel conductors of infinite length and one metre apart in empty space, causes each conductor to experience a force of exactly 2×10^{-7} newton per metre of length.

Derived S.I. Units

Physical quantity	Name of unit	Symbol for unit
(a) With special names		
force	newton	$N = kg\ m\ s^{-2}$
energy	joule	$J = kg\ m^2\ s^{-2}$
power	watt	$W = kg\ m^2\ s^{-3}$
electric charge	coulomb	$C = A\ s$
electric potential difference	volt	$V = J\ A^{-1}\ s^{-1}$
magnetic flux	weber	$Wb = V\ s$
magnetic induction	tesla	$T = V\ s\ m^{-2}$
(magnetic flux density)		$= Wb\ m^{-2}$
frequency	hertz	$Hz = s^{-1}$

Physical quantity	Name of unit	Symbol for unit
(b) Others, without special names		
velocity	metre per second	$m\ s^{-1}$
acceleration	metre per second^2	$m\ s^{-2}$
electric intensity	volt per metre	$V\ m^{-1}$

Submultiple	Prefix	Symbol	Multiple	Prefix	Symbol
10^{-3}	milli	m	10^3	kilo	k
10^{-6}	micro	μ	10^6	mega	M
10^{-9}	nano	n	10^9	giga	G
10^{-12}	pico	p	10^{12}	tera	T
10^{-15}	femto	f			
10^{-18}	atto	a			

Examples

1 nm is one nanometre, which is 10^{-9} metre.
1 km is one kilometre, which is 10^3 metre.

Non-S.I. Units

The electron volt (eV) is $1 \cdot 60206 \times 10^{-19}$ J (see p. 19).
$$1 \text{ MeV} = 10^6 \text{ eV}; \quad 1 \text{ GeV} = 10^9 \text{ eV}$$
The ångström (Å) is 10^{-10} m

Symbols

Symbols for units are printed as above, in Roman (ordinary) type, without a full-stop. The full names of units are written without a capital (e.g. joule), and in formal statements the plural ' s ' is not used.

Symbols printed in italics represent values of physical quantities. Heavy type denotes vectors. Thus, u stands for the magnitude of the vector **u**.

(a) *Italic symbols*

a acceleration.

B magnetic induction measured in tesla. It can be defined in terms of the magnetic force on a moving charge or on a length of wire carrying a current. Some engineers prefer to define B in terms of induced emf, calling it magnetic flux density, and measuring it in Wb m^{-2}.

c speed of light in empty space. Experimentally, $c = 2 \cdot 99793 \times 10^8$ m s^{-1}.

E electric field intensity vector, measured in V m^{-1}.

E total energy of a particle. It is equal to the kinetic energy T plus the rest mass energy $m_0 c^2$.

f force.

g acceleration due to gravity.

G gravitational constant. Experimentally, $G = 6 \cdot 67 \times 10^{-11}$ N m^2 kg^{-2}.

K the expression $[(1 + v/c)/(1 - v/c)]^{1/2}$.

l length.

l_0 proper length.

m relativistic mass $= m_0/(1 - u^2/c^2)^{1/2}$.

m_0	rest mass.
n	frequency.
n	refractive index.
p	momentum; p_x, p_y, p_z, momentum components.
q	charge.
r	distance from a point.
t	time.
T	kinetic energy.
T_0	mean life.
$T_{1/2}$	half life.
u	velocity of a particle; u_x, u_y, u_z, velocity components.
v	velocity of the inertial reference frame Σ' relative to the inertial reference frame Σ.
w	product (velocity of light $\times$ time); $w = ct$.
x, y, z,	co-ordinates.

(b) *Greek symbols*:

α	alpha. Used to represent an angle.
β	beta.
γ	gamma. Used as an abbreviation for $1/(1 - v^2/c^2)^{1/2}$.
δ	delta. Used to represent an increment.
Δ	capital delta. Used to represent an increment.
ϵ	epsilon.
ϵ_0	electric space constant. Experimentally $\epsilon_0 = 8\cdot854 \times 10^{-12}$ farad per metre.
θ	theta. Used to represent an angle.
λ	lambda. It is used in the text (a) to represent the wavelength of light, (b) to represent the radioactive decay constant, and (c) to represent geographic latitude.
μ	mu.
μ_0	magnetic space constant. It follows from the definition of the ampere that $\mu_0 = 4\pi \times 10^{-7}$ henry per metre.
ν	nu. Used to represent the frequency of light.
π	pi.
ρ	rho. Used to represent radius of curvature.
Σ	capital sigma. Used to label inertial reference frames.
ϕ	phi. Used to represent angles.
ω	omega. Used to represent angular velocity.

Primes

We shall be considering an inertial reference frame Σ, and a second inertial reference frame Σ', moving with uniform velocity v along the x-axis relative to Σ. All measurements in the inertial reference frame Σ are written without primes, as x, y, z; u_x, u_y, u_z etc. All measurements made in the inertial reference frame Σ' are written with primes, as x', y', z'; u_x'; u_y'; u_z' etc.

CONTENTS

x

CHAPTER 1
Newtonian mechanics and the principle of relativity

1.1 *Introduction*

MEASUREMENTS of the positions of events must be made with respect to an origin and axes, that is, a co-ordinate system, which we shall call a *reference frame*. The velocity of a particle relative to a reference frame is determined by measuring how far the particle moves per unit time relative to that reference frame. If one were in a terrestrial laboratory, it would be convenient to use an origin and co-ordinate system at rest in that laboratory. However, if one were inside an ocean liner, it would be more convenient to use a co-ordinate system (reference frame), at rest relative to the ocean liner, to determine the positions of events occurring inside the ocean liner. In this chapter it will be illustrated how Newton's laws of motion can be applied inside the ocean liner, if it is moving with uniform velocity relative to the earth. Newton's laws, in modern terms, read:

(i) Every body continues in its state of rest or of uniform motion in a straight line unless it is compelled to change that state by an external impressed force.

(ii) The rate of change of momentum is proportional to the impressed force, and takes place in the direction of the force.

(iii) Action and reaction are equal and opposite. That is, between two bodies the force exerted by one on the other is equal in magnitude to the force it experiences from the other and in the opposite direction.

The first law was first stated clearly by Galileo, and is known as the *principle of inertia*. A reference frame in which the first law holds is called an *inertial reference frame* or a *Galilean reference frame*. In an inertial reference frame, according to Newton's second law, if the mass of a body is constant:

$$f = \frac{\mathrm{d}(mu)}{\mathrm{d}t} = m\frac{\mathrm{d}u}{\mathrm{d}t} = ma, \qquad (1.1)$$

where f is the force acting on a body of mass m having velocity u, momentum mu and acceleration a. The reader should remember that Newton's laws are developed from experiments with big bodies, such as billiards balls, moving at speeds very much less than the speed of light. For example, a speed of 100 kilometre per hour, which is 28 metre per second, is less than $10^{-7}c$, where $c = 3 \times 10^8$ metre per second is the speed of light. Later in this book we shall consider

1   1

particles moving with speeds comparable to c. This goes beyond the range of our normal everyday experiences, and the reader should approach special relativity in Chapter 4 with an open mind, realizing that it will require the development of new concepts on the part of the reader.

1.2 *The principle of relativity*

Consider a large ocean liner moving out to sea with *uniform* velocity on a calm day. It is quite possible for the passengers to play table tennis in the games room in such conditions. There is no need for the players to stop and wonder between each stroke which way the ship is going. They can play their normal game, just as if they were playing on dry land, and they can ignore the motion of the ship. Of course, in rough seas the sudden accelerations of the ship would affect their normal game of table tennis, and they would have to try and allow for the accelerations of the ship. The discussion of accelerating reference frames will be deferred until Chapter 6.

On the basis of mechanical experiments on large bodies carried out inside a ship moving with *uniform* velocity, one would conclude that Newton's laws held to a very good approximation relative to the ship. Without looking at anything external to the ship one could not say, on the basis of these experiments, whether the ship was even moving. If one were told that it was moving, one could not determine the speed of the ship without looking at something external to the ship. This is an example of the *principle of relativity*, according to which the laws of physics are the same in all inertial reference frames. Though the laws are the same, it does not mean that the measurements of particular events are the same in all inertial reference frames. For example, if a ball is rolling at a speed of, say, 10 metre per second relative to the ship in the direction of motion of the ship, then, according to Newtonian mechanics, the speed of the ball relative to the earth should be $(10 + v)$, where v is the speed of the ship relative to the earth. The reader will probably accept it as quite natural that the speed of the ball will be different, when the standard of rest is changed.

We will now proceed to see how the values of various other quantities, measured relative to one inertial reference frame, can be changed (or transformed) into the numerical values measured relative to an inertial reference frame moving with uniform velocity relative to the first. Apart from its usefulness in practical cases, this procedure illustrates some of the assumptions made in Newtonian mechanics, which are not always apparent when Newtonian mechanics is developed for the first time.

1.3 *The Galilean transformations*

Still considering the ocean liner moving with uniform velocity v, we will now solve the same problem, namely that of a falling body, relative to the moving ship and the earth.   Let a mass m be dropped from the top of the ship's mast as shown in fig. 1.1.   Let the laboratory system, that is a reference frame which is at rest relative to the earth be denoted by Σ, *where the symbol Σ is the Greek capital letter sigma.* Let a reference frame in which the ship is at rest be denoted by Σ'. Choose a point O′ at the top of the ship's mast, level with the point from which the mass m is dropped, as the origin of Σ', as shown in fig. 1.1 *a*.   Let the mass m be dropped from rest relative to the ship, at the time $t=0$, from a point a distance X_0 from O′, as shown in fig. 1.1 *a*.   Let O the origin of Σ coincide with O′ the origin of Σ' at

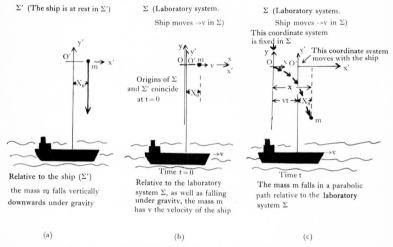

Figure 1.1.   A mass m is dropped from the mast of a moving ship.   (*a*) Relative to the ship the mass m falls vertically downwards in a straight line.   (*b*) and (*c*) The path of the falling body is a parabola relative to the earth.

the time $t=0$, when the mass m is dropped, as shown in fig. 1.1 *b*. Let the x axis of Σ and the x' axis of Σ' be parallel to v, the velocity of the ship relative to the earth, so that Σ' is moving with uniform velocity v along the x axis of Σ.   Let the y axis of Σ and the y' axis of Σ' be in the vertical direction, as shown in fig. 1.1 *b*.   The mass m is dropped from a point having co-ordinates

$$x'=X_0; \quad y'=0; \quad t=0$$

relative to the ship (Σ') as shown in fig. 1.1 *a*.   According to Newton's law of universal gravitation, relative to the ship (Σ') the force of gravitational attraction on the mass m is:

$$f'=G\frac{mM}{r^2}, \tag{1.2}$$

3

where M is the mass of the earth, r is the distance of the mass m from the centre of the earth and G is the gravitational constant. Applying Newton's second law, equation (1.1), relative to the ship (Σ'), we have, whilst the mass m is falling under gravity:

$$f' = ma',\qquad\qquad (1.3)$$

or equating equations (1.2) and (1.3) and cancelling m:

$$a' = \frac{GM}{r^2},\qquad\qquad (1.4)$$

where a' is the acceleration due to gravity relative to the ship (Σ'). At the time $t=0$, relative to the ship (Σ'), the position of the mass m is:

$$x' = X_0; \quad y' = 0.$$

Since the body is dropped from rest relative to the ship, it has no velocity in the x' direction relative to the ship (Σ') in fig. 1.1 a. In the $+y'$ direction it has an acceleration of $-a'$. Hence after a time t, relative to the ship (Σ'), the position of the mass m is:

$$x' = X_0,\qquad\qquad (1.5)$$

$$y' = -\frac{1}{2}a't^2 = -\frac{1}{2}\left(\frac{GM}{r^2}\right)t^2.\qquad\qquad (1.6)$$

(The variation of r with the height of the mass m above sea level is being ignored.) Thus, according to Newtonian mechanics, relative to the ship (Σ'), the mass m falls vertically downwards in a straight line with the acceleration due to gravity which is given by equation (1.4).

The same problem will now be solved relative to the laboratory system (Σ), in which the earth is at rest and the ship is moving with uniform velocity v. According to Newton's law of universal gravitation, the force of gravitational attraction on the mass m, measured relative to the earth (Σ) is:

$$f = G\,\frac{mM}{r^2}.\qquad\qquad (1.7)$$

When the body is falling, from Newton's second law, we have:

$$f = ma.\qquad\qquad (1.8)$$

Equating equations (1.7) and (1.8) we have for the acceleration of the falling body:

$$a = \frac{GM}{r^2}.\qquad\qquad (1.9)$$

Since the ship is moving relative to the earth (Σ), when the mass m is dropped at the time $t=0$, the mass m starts with a velocity v in

4

the $+x$ direction relative to the earth, as shown in fig. 1.1 b. Since the acceleration of the mass m is in the vertical (or $-y$) direction, the velocity of the mass m in the x direction remains constant. At $t=0$ the co-ordinates and velocity of the mass m, relative to the earth (Σ), are:

$$x=X_0; \quad y=0; \quad u_x=v; \quad u_y=0.$$

(Suffixes x, y and z are used to denote the components of vector quantities such as the velocity $\mathbf{u}$ relative to the x, y and z axes respectively. Heavy type will be used to denote vectors.)   After a time t, the position of the mass m relative to the earth (Σ) is:

$$x=X_0+vt, \tag{1.10}$$

$$y=-\frac{1}{2}\,at^2=-\frac{1}{2}\left(\frac{GM}{r^2}\right)t^2. \tag{1.11}$$

Relative to the earth, the falling mass m moves in a parabolic path as illustrated in fig. 1.1 c.   Notice it was *assumed* that the force, given by Newton's law of gravitation, was the same relative to both the ship (Σ') and the laboratory system (Σ).   This illustrates how it is generally *assumed* in Newtonian mechanics that the force acting on a body is an invariant, that is, has the same numerical value in all inertial reference frames.   Notice also the same value was used for the mass of the body at all times in Σ and Σ'.   This illustrates how in Newtonian mechanics it is *assumed* that the mass of a body is independent of its velocity, and has the same numerical value in all inertial reference frames.   It was also *assumed* that time is absolute, the same symbol t being used to denote the time measured relative to both the earth (Σ) and the ship (Σ').

It will now be shown how to convert the results measured in one inertial reference frame into the results appropriate to the other inertial reference frame.   Consider the measurement of the position of the falling mass m after a time t, using rulers parallel to the x and y axes respectively, as shown in fig. 1.2.   It is *assumed* in Newtonian mechanics that time is absolute.   If time were absolute, an observer at rest relative to the earth (Σ) and an observer at rest relative to the ship (Σ') should agree that at the time t the y axis of Σ, the y' axis of Σ' and the mass m coincided with the marks X_1, X_2 and X_3 on the ruler, which is parallel to the x and x' axes, as shown in fig. 1.2.   (The ruler may be moving relative to both Σ and Σ'.)   Hence, according to Newtonian mechanics, observers at rest relative to the earth and the ship should agree that in fig. 1.2:

$$X_1X_3=X_1X_2+X_2X_3.$$

Since the speed of the ship is v, in a time t the ship goes a distance $X_1X_2=vt$ relative to the laboratory system (Σ).   Since $X_1X_3=x$ and $X_2X_3=x'$, as shown in fig. 1.2, then

$$x=x'+vt.$$

5

Similarly, according to Newtonian mechanics observers at rest relative to the earth (Σ) and the ship (Σ') should agree that the x and x' axes of Σ and Σ' coincide with the point Y_1 on the vertical ruler at the same time t that the mass m coincides with the position Y_2. Hence the observers should agree on the vertical distance fallen by the body, so that $y=y'$. Hence, if it is *assumed* that time is absolute, that is

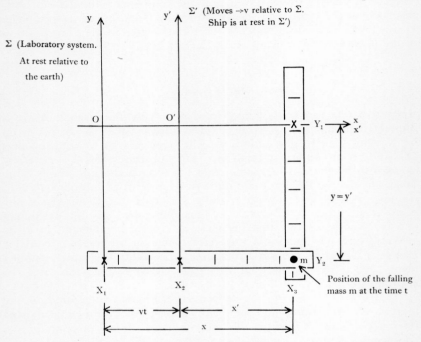

Figure 1.2.   The measurement of the x and x' and the y and y' co-ordinates of the falling mass m relative to Σ and Σ' respectively.

that the observers on the earth (Σ) and the ship (Σ') agree on the times of all events, then the co-ordinates x, y, z of an event at a time t in Σ are related to the co-ordinates of the *same* event at x', y', z' at a time t' in Σ' by the following relations:

$$x=x'+vt'; \quad x'=x-vt, \qquad (1.12)$$

$$y=y'; \qquad y'=y, \qquad (1.13)$$

$$z=z'; \qquad z'=z, \qquad (1.14)$$

$$t=t'; \qquad t'=t. \qquad (1.15)$$

These are the Galilean co-ordinate transformations.   The same units of length, time and mass must be used in both Σ and Σ'.   For example,

6

the observers could agree to define the metre as 1 650 763·73 wavelengths of the orange–red line of krypton–86. They could define the
second in terms of the frequency of some atomic or nuclear transition.
The kilogramme could be defined in terms of the rest mass of the
proton. For example, they could assume, by definition, that the rest
mass of the proton was $1·67239 \times 10^{-27}$ kilogramme. For electrical
measurements they could define the coulomb in terms of the charge
on a proton.

Substituting for x' from equation (1.12) into equation (1.5) gives:

$$x' = X_0 = x - vt$$

or

$$x = X_0 + vt. \tag{1.16}$$

Substituting for y' from equation (1.13) into equation (1.6), we find:

$$y' = y = -\frac{1}{2} a't^2 = -\frac{1}{2} \left(\frac{GM}{r^2}\right)t^2. \tag{1.17}$$

Equations (1.16) and (1.10) are the same and equations (1.17) and
(1.11) are the same, showing how the Galilean transformations can be
used to convert measurements from one inertial reference frame to the
other. Similarly, equations (1.12) and (1.13) can be used to convert
equations (1.10) and (1.11) into equations (1.5) and (1.6) respectively.

Differentiating equations (1.12), (1.13) and (1.14) with respect
to time, we have, since v, the speed of the ship, is assumed to be a
constant:

$$\frac{dx}{dt} = \frac{dx'}{dt} + v; \quad \frac{dy}{dt} = \frac{dy'}{dt}; \quad \frac{dz}{dt} = \frac{dz'}{dt}.$$

Hence

$$u_x = u_x' + v, \tag{1.18}$$

$$u_y = u_y', \tag{1.19}$$

$$u_z = u_z'. \tag{1.20}$$

These are the Galilean velocity transformations.

1.4 *Discussion*

Accelerating reference frames will not be considered until Chapter 6,
and until then the discussion will be restricted to the special (or
restricted) theory of relativity, which is only applicable to inertial
reference frames. In Newtonian mechanics an inertial reference frame
is defined as a reference frame in which Newton's first law (the principle
of inertia) is valid, that is a reference frame in which a body, which
is not acted upon by any forces (for example, a body far away from

7

any bodies capable of exerting forces), moves in a straight line with uniform velocity. The same definition of an inertial reference frame is used in the theory of special relativity. Due to the rotation of the earth, strictly, the *laboratory system*, that is a reference frame at rest relative to the earth, is *not* an inertial reference frame, and the effects associated with the earth's rotation are sometimes important, for example, in long-range naval gunnery and Foucault's pendulum experiment. In these cases we get a better approximation to an inertial reference frame by taking a reference frame at rest relative to the solar system or the fixed stars. However, the angular velocity of rotation of the earth is only $7 \cdot 3 \times 10^{-5}$ radian per second (the earth turns through $360°$ in 24 hours), so that the effects associated with the earth's rotation about its axis are generally very small. Effects due to the earth's rotation about the sun are much smaller still. (The earth goes through $360°$ around the sun in a year.) Effects associated with phenomena such as the rotation of the galaxy are very much smaller again. For most low speed mechanical experiments the reference frame in which the earth is at rest, that is the laboratory system, is generally a satisfactory approximation to an inertial reference frame, provided that gravity is treated as giving an ' applied force '. The theory of special relativity is generally applied to optics, electricity and magnetism, and atomic and nuclear physics. The effect of the earth's rotation in all these cases is usually negligible, and the laboratory system is almost invariably a satisfactory approximation to an inertial reference frame in which to apply the laws of optics, electricity and magnetism and atomic and nuclear physics.

According to the principle of relativity, one can apply the laws of physics in any inertial reference frame. In order to remove the mental bias in favour of the laboratory system, we shall, in Chapter 4, generally consider two spaceships coasting in outer space with uniform velocities relative to the fixed stars. All that the astronauts on these spaceships can say is that they are moving with uniform velocity relative to each other. Observers at rest in an inertial reference frame will be called inertial observers.

Problems

1.1. If you were locked in the hold of a ship, how would you prove that the ship was (*a*) moving with uniform velocity, (*b*) accelerating relative to the earth?

1.2. A ship is travelling due east at a speed of 25 m s^{-1}.
   (*a*) A ball is rolled due north on the deck of the ship at a speed of 5 m s^{-1} relative to the ship. What is its velocity relative to the earth?
   (*b*) If the ball is rolled $30°$ east of north at a speed of 5 m s^{-1} relative to the ship, what is its speed relative to the earth?

8

1.3. A stone is dropped from rest from the mast of a ship moving with a velocity of 15 m s^{-1} relative to the earth. Choose the origins of Σ (the laboratory frame) and Σ' (the co-ordinate system in which the ship is at rest) such that the origins coincide with the stone at the instant $t=0$, when the stone is dropped. If the acceleration due to gravity is $9\cdot8$ m s^{-2}, find the position of the stone (*a*) relative to the ship, (*b*) relative to the laboratory frame after 2 s. Show how the results can be related using the Galilean transformations. The ship is moving parallel to the x axis of Σ.

CHAPTER 2
high energy physics: relativistic mechanics

2.1 *Introduction*
IN this chapter it will be shown that at very high speeds close to the speed of light, Newtonian mechanics is completely inadequate. The fastest speeds of aeroplanes at present are about 3000 kilometre per hour, which is only about $3 \times 10^{-6}c$, where c is the speed of light. The only particles that can be accelerated up to speeds comparable to the speed of light are atomic particles, such as protons and electrons, so that small size generally goes with high speed. Both the high speeds and the small sizes involved in high energy physics go beyond the realm of direct perception and beyond our 'common sense', which is almost invariably based on the concepts of Newtonian mechanics. The reader must approach the development of high energy physics and special relativity with an open mind, and be guided by experiment rather than rely on 'sound common sense'.

2.2 *Variation of mass with velocity*
In 1908 Bucherer measured the ratio of charge to mass (e/m) for β-ray electrons and showed that at high speeds, comparable to the speed of light, the masses of the electrons depended on the speeds of the electrons. Before giving an account of Bucherer's experiment, a brief review will be given of the theory of the deflection of moving charges in electric and magnetic fields.

Consider a positive charge $+q$ coulomb moving between the plates of a parallel plate capacitor as shown in fig. 2.1 *a*. If a potential difference V is applied across the plates of the capacitor, which are a distance d metre apart, there is an electric field of strength E equal to V/d volt per metre across the capacitor, as shown in fig. 2.1 *a*. Experiments have shown that the electric force (denoted f_{elec}) on the charge $+q$ is:

$$f_{\text{elec}} = +qE \text{ newton.} \tag{2.1}$$

For a positive charge, this force acts in the direction of the electric field, and leads to the deflection of the moving electric charge, as shown in fig. 2.1 *a*. It is assumed that the electric force on a moving charge is independent of the velocity of the charge.

The motion of a moving charge in a magnetic field will now be considered. Let a charge $+q$ move with velocity u at an angle α to the direction of the magnetic induction B, as shown in fig. 2.1 *b*. The

10

magnitude of the magnetic force (denoted f_{mag}) is given by:

$$f_{mag} = quB \sin \alpha. \qquad (2.2)$$

If q is in coulomb, u in metre per second and B in tesla (weber per square metre, volt-second per square metre), then f_{mag} is in newton. The direction of this magnetic force is given by the left-hand motor

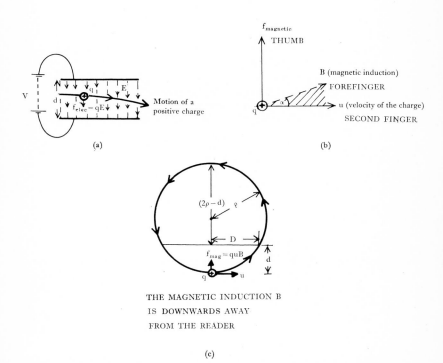

(a)

(b)

THE MAGNETIC INDUCTION B

IS DOWNWARDS AWAY

FROM THE READER

(c)

Figure 2.1. (*a*) Deflection of a moving charge in an electric field. (*b*) The left-hand motor rule. (*c*) Deflection of a moving charge in a magnetic field.

rule. If the forefinger of the left-hand points in the direction of the magnetic induction B, the second finger in the direction of the velocity of the charge u (for a positive charge), then the thumb of the left hand points in the direction of the magnetic force acting on the moving charge, as shown in fig. 2.1 *b*. This magnetic force is perpendicular to the plane containing u and B, as shown in fig. 2.1 *b*.

Consider the motion of a charge $+q$ in a uniform magnetic induction field of strength B tesla. Let the velocity of the charge $+q$ be u

11

metre per second in a direction perpendicular to the direction of the magnetic induction, as shown in fig. 2.1 c. The direction of the magnetic induction is perpendicular to the paper, in the direction away from the reader in fig. 2.1 c. Since in this case u is perpendicular to B, $\alpha = 90°$, $\sin \alpha = 1$ and equation (2.2) becomes:

$$f_{\text{mag}} = quB.$$

The direction of the magnetic force is shown in fig. 2.1 c. Since this force is always perpendicular to the velocity of the charge $+q$, though the magnetic force leads to a change in the direction of u, it does not give rise to a change in the magnitude of u. Hence the charge $+q$ in fig. 2.1 c moves with uniform speed u in a circle in the plane of the paper in fig. 2.1 c. If the radius of the circle is ρ, from the theory of circular motion, the centripetal acceleration is u^2/ρ and the force necessary to give such a centripetal acceleration is mu^2/ρ. This centripetal force is the magnetic force acting on the charge. Hence

$$f_{\text{mag}} = quB = \frac{mu^2}{\rho} \qquad (2.3)$$

or

$$mu = qB\rho, \qquad (2.4)$$

$$\rho = \frac{mu}{qB}. \qquad (2.5)$$

The time T for the charge $+q$ to go around one complete circle in fig. 2.1 c is equal to $2\pi\rho$, the circumference of the circle, divided by u the speed of the particle, that is:

$$T = \frac{2\pi\rho}{u} = \frac{2\pi m}{qB}. \qquad (2.6)$$

In fig. 2.1 c, from the law of intersecting chords:

$$D^2 = d(2\rho - d),$$

where D and d are illustrated in fig. 2.1 c. Hence, using equation (2.5):

$$\rho = \frac{(D^2 + d^2)}{2d} = \frac{mu}{qB}. \qquad (2.7)$$

When a charge moves in both an electric and a magnetic field there is both an electric force given by equation (2.1) and a magnetic force given by equation (2.2) acting on the moving charge. The total force acting on the moving charge is the vector sum of these forces:

$$\mathbf{f}_{\text{total}} = \mathbf{f}_{\text{elec}} + \mathbf{f}_{\text{mag}}. \qquad (2.8)$$

12

The total force given by equation (2.8) is generally known as Lorentz's expression for the total force acting on a moving charge, or sometimes just the *Lorentz force law*.

The principle of Bucherer's experiment is illustrated in fig. 2.2; β-ray electrons emitted by a radium source were passed between the plates of a large parallel plate capacitor. The diameters of the capacitor plates were very much bigger than their separation. The apparatus was placed inside a vacuum. A potential difference was applied to the plates of the capacitor, so as to produce an electric field in the negative y direction in fig. 2.2. Since the charge on the electron is negative, $q = -e$, then the electric force eE acting on an electron moving

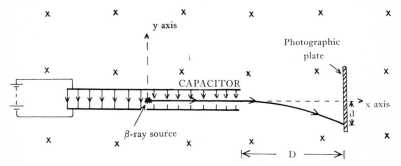

Figure 2.2. Simplified form of Bucherer's experiment. The capacitor acts as a velocity selector. After emerging from between the plates of the capacitor the β-rays are deflected in the magnetic field and detected by a photographic plate. The magnetic induction B is downwards away from the reader.

between the plates of the capacitor is in the opposite direction to the direction of the electric field, that is in the $+y$ direction in fig. 2.2. A magnetic field was applied in the negative z direction, that is perpendicular to the paper away from the reader in fig. 2.2. Consider a β-ray electron moving along the $+x$ axis. According to equation (2.2), since u the velocity of the electron is perpendicular to the magnetic induction B, $\alpha = 90°$ and the magnetic force on the electron of charge $-e$, is euB. According to the left-hand motor rule this force is in the negative y direction. This is opposite in direction to the electric force eE on the electron. If the electric and magnetic forces acting on the electron are not exactly equal and opposite, the electron will be deflected in either the positive or the negative y direction, and will not emerge from between the plates of the capacitor. If the electric and magnetic forces are equal and opposite in directions, an electron moving along the x axis will not be deflected and will emerge from between the plates of the capacitor. For this case the electric and magnetic forces are equal, that is $euB = eE$, so that

$$u = E/B. \tag{2.9}$$

13

The capacitor acted as a velocity selector in Bucherer's experiment, selecting β-ray electrons of velocities given by equation (2.9). In Bucherer's experiment there was no electric field outside the capacitor plates, and after emerging from between the plates of the capacitor, the electrons moved in circular orbits in the magnetic field, before striking the photographic plate in fig. 2.2. From equation (2.7), if the deflection of the electrons is d, as shown in fig. 2.2, we have:

$$\frac{mu}{Be} = \frac{(D^2 + d^2)}{2d}.$$

From equation (2.9), $u = E/B$, so that after rearranging we have:

$$\frac{e}{m} = \frac{2d}{(D^2 + d^2)} \frac{E}{B^2}.$$

In S.I. units e/m is in coulomb per kilogramme ($C\ kg^{-1}$). By measuring d, D, E and B, e/m can be calculated. Some of Bucherer's results are given in Table 2.1.

u/c	e/m	$\dfrac{e}{m_0} = \dfrac{e}{m\sqrt{(1-u^2/c^2)}}$
0·3173	$1·661 \times 10^{11}\ C\ kg^{-1}$	$1·752 \times 10^{11}\ C\ kg^{-1}$
0·3787	$1·630 \times 10^{11}$	$1·761 \times 10^{11}$
0·4281	$1·590 \times 10^{11}$	$1·760 \times 10^{11}$
0·5154	$1·511 \times 10^{11}$	$1·763 \times 10^{11}$
0·6870	$1·283 \times 10^{11}$	$1·767 \times 10^{11}$

Table 2.1.

It can be seen that the experimental values of e/m depend on the speeds of the electrons. However, if one assumes that

$$m = \frac{m_0}{\sqrt{(1-u^2/c^2)}}, \tag{2.10}$$

where u is the speed of the β-ray electron and c is the speed of light, and if one calculates

$$\frac{e}{m_0} = \frac{e}{m\sqrt{(1-u^2/c^2)}},$$

then the calculated values of e/m_0 given in Table 2.1 are remarkably constant. They are as good a set of results as the reader is ever likely to obtain in his own laboratory work. In the spirit in which physical laws are 'established by experiment' in elementary practical courses, we will conclude from Bucherer's *experiment* that equation (2.10) is established *by experiment*. The quantity m in equation (2.10), which appears also in equations (2.3), (2.4), (2.5), (2.6) and (2.7), is generally called the *relativistic mass* or just the *mass* of the particle. The quantity

14

m_0, which is the value of m when $u = 0$, is called the *rest mass* or *proper mass* of the particle. It can be seen that as the velocity of the particle increases, according to equation (2.10) the mass of the particle increases.

Notice we assumed that the charge $-e$ on the electron was independent of its velocity. Instead of saying that mass varied according to equation (2.10), we might be tempted to say that the charge q on a particle varied according to the equation:

$$q = q_0 \sqrt{(1 - u^2/c^2)}, \qquad (2.11)$$

where u was the velocity of the charge, q_0 the value of the charge when it was at rest, and that the mass m was invariant. Such assumptions would account for the results given in Table 2.1. There is, however, independent evidence in favour of the principle of constant electric charge. For example, if the charge on a particle did vary with velocity according to equation (2.11), then hydrogen atoms and molecules would not be electrically neutral, since the negative electrons are moving in orbits around the atomic nuclei in hydrogen atoms and molecules, and on average are moving faster than the positive nuclei (protons in this case) relative to the laboratory. If the charge did vary with velocity, hydrogen molecules should be deflected in electric fields, e.g. of the type shown in fig. 2.1 *a*. In 1960 King showed that the charges on the electrons and the protons in hydrogen molecules were numerically equal to within one part in 10^{20}. We therefore conclude that the charge on a particle is independent of its velocity and that the mass of a particle varies with the particle's velocity, according to equation (2.10).

To simplify the mathematics, we will sometimes make the trigonometrical substitution:

$$u = c \sin \theta \ \text{ or } \ u/c = \sin \theta, \qquad (2.12)$$

where u is the velocity of the particle and c is the velocity of light. Substituting in equation (2.10):

$$m = \frac{m_0}{\sqrt{(1 - \sin^2 \theta)}} = \frac{m_0}{\sqrt{(\cos^2 \theta)}} = m_0 \sec \theta. \qquad (2.13)$$

The variation of mass with velocity can be shown by plotting $m/m_0 = \sec \theta$ against $u/c = \sin \theta$ as shown in fig. 2.3. As $u \to c$, m/m_0 tends to infinity. For normal laboratory speeds the variation of mass with velocity is negligible. Consider a train going at 100 kilometre per hour, which corresponds to $u/c \simeq 10^{-7}$. In this case:

$$m = \frac{m_0}{\sqrt{(1 - 10^{-14})}} = m_0 (1 - 10^{-14})^{-1/2}.$$

According to the binomial theorem, if $x \ll 1$:

$$(1 + x)^n \simeq 1 + nx.$$

15

Putting $x = -10^{-14}$ and $n = -\frac{1}{2}$, we have:

$$m = m_0(1 - 10^{-14})^{-1/2} = m_0(1 + \frac{1}{2} \, 10^{-14}),$$

$$m = 1 \cdot 000\ 000\ 000\ 000\ 005\ m_0.$$

It can be seen that one can ignore the variation of mass with velocity in our normal everyday lives.

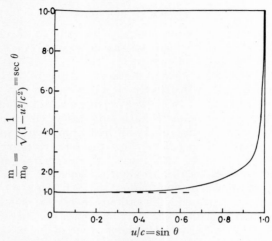

Figure 2.3. A plot of sec θ against sin θ. This is equivalent to plotting $m/m_0 = 1/\sqrt{(1 - u^2/c^2)}$ against u/c.

2.3 *The laws of high speed particles*

The changes in other mechanical quantities, arising from the variation of mass with velocity, given by equation (2.10), will now be discussed. In high energy physics the momentum of a particle is denoted by **p** and is put equal to the product of the relativistic mass m and the velocity of the particle **u**. That is:

$$\mathbf{p} = m\mathbf{u} = \frac{m_0 \mathbf{u}}{\sqrt{(1 - u^2/c^2)}}. \tag{2.14}$$

Writing equation (2.14) in components we have:

$$p_x = \frac{m_0 u_x}{\sqrt{(1 - u^2/c^2)}}; \quad p_y = \frac{m_0 u_y}{\sqrt{(1 - u^2/c^2)}}; \quad p_z = \frac{m_0 u_z}{\sqrt{(1 - u^2/c^2)}}. \tag{2.15}$$

Putting $u/c = \sin\theta$ in equation (2.14), we have:

$$p = \frac{m_0 u}{\sqrt{(1 - u^2/c^2)}} = \frac{m_0 c \sin\theta}{\sqrt{(1 - \sin^2\theta)}} = \frac{m_0 c \sin\theta}{\cos\theta}$$

$$p = m_0 c \tan\theta. \tag{2.16}$$

16

In high speed mechanics the force acting on a particle is put equal to the rate of change of the momentum of the particle, that is:

$$\mathbf{f} = \frac{\mathrm{d}\mathbf{p}}{\mathrm{d}t} = \frac{\mathrm{d}}{\mathrm{d}t}(m\mathbf{u}) = \frac{\mathrm{d}}{\mathrm{d}t}\left(\frac{m_0\mathbf{u}}{\sqrt{(1 - u^2/c^2)}}\right). \tag{2.17}$$

Due to the variation of the mass of the particle with its velocity, we have:

$$\mathbf{f} = m\frac{\mathrm{d}\mathbf{u}}{\mathrm{d}t} + \mathbf{u}\frac{\mathrm{d}m}{\mathrm{d}t} = m\mathbf{a} + \mathbf{u}\frac{\mathrm{d}m}{\mathrm{d}t}$$

If $\mathbf{a}$, the acceleration of the particle, leads to a change in its speed, its mass also changes and $\mathrm{d}m/\mathrm{d}t$ is finite. Hence the relation force equals mass times acceleration is not generally valid for high speed particles.

The total force acting on a charge q moving with velocity $\mathbf{u}$ in an electric field $\mathbf{E}$ and a magnetic field $\mathbf{B}$ is given by Lorentz's expression for the force on a moving charge, namely equation (2.8), which, using equation (2.17), becomes:

$$\mathbf{f}_{\text{total}} = \mathbf{f}_{\text{elec}} + \mathbf{f}_{\text{mag}} = \frac{\mathrm{d}}{\mathrm{d}t}\left(\frac{m_0\mathbf{u}}{\sqrt{(1 - u^2/c^2)}}\right), \tag{2.18}$$

where, according to equations (2.1) and (2.2):

$$\mathbf{f}_{\text{elec}} = q\mathbf{E} \tag{2.19}$$

and

$$f_{\text{mag}} = quB \sin \alpha. \tag{2.20}$$

The direction of $\mathbf{f}_{\text{mag}}$ is given by the left-hand motor rule. The angle α is the angle between u and B. It will be illustrated in § 2.6 how equation (2.18) can be used to design the high energy accelerators used in high energy nuclear physics.

If the total force f acting on a particle gives rise to a displacement $\mathrm{d}l$ in the direction of the force, the work done by the force is $f\,\mathrm{d}l$. If it is assumed that all the work done goes into increasing the kinetic energy of the particle, then the increase in the kinetic energy of the particle is given by:

$$\mathrm{d}T = f\,\mathrm{d}l = f\frac{\mathrm{d}l}{\mathrm{d}t}\,\mathrm{d}t = f u\,\mathrm{d}t, \tag{2.21}$$

where the symbol T is used to denote the kinetic energy of the particle, and $u = \mathrm{d}l/\mathrm{d}t$ is the velocity of the particle. From equations (2.17) and (2.16):

$$f = \frac{\mathrm{d}p}{\mathrm{d}t} = \frac{\mathrm{d}}{\mathrm{d}t}(mu) = \frac{\mathrm{d}}{\mathrm{d}t}(m_0c \tan \theta) = m_0c \sec^2 \theta \frac{\mathrm{d}\theta}{\mathrm{d}t},$$

17

and from equation (2.12):

$$u = c \sin \theta.$$

Substituting in equation (2.21), we have:

$$dT = m_0 c \sec^2 \theta \frac{d\theta}{dt} c \sin \theta \, dt = m_0 c^2 \frac{\sin \theta}{\cos^2 \theta} \, d\theta.$$

Assuming the particle starts from rest, when $T=0$ and $\theta = 0$, we have:

$$\int_0^T dT = m_0 c^2 \int_0^\theta \frac{\sin \theta}{\cos^2 \theta} \, d\theta = -m_0 c^2 \int_0^\theta \frac{d(\cos \theta)}{\cos^2 \theta}, \qquad (2.22)$$

$$[T]_0^T = m_0 c^2 \left[\frac{1}{\cos \theta} \right]_0^\theta = m_0 c^2 [\sec \theta - 1].$$

Now

$$\sec \theta = \frac{1}{\cos \theta} = \frac{1}{\sqrt{(1 - \sin^2 \theta)}} = \frac{1}{\sqrt{(1 - u^2/c^2)}}. \qquad (2.23)$$

Hence the kinetic energy of the particle is given by:

$$T = m_0 c^2 [\sec \theta - 1] = m_0 c^2 \left[\frac{1}{\sqrt{(1 - u^2/c^2)}} - 1 \right]. \qquad (2.24)$$

Expanding the right-hand side of equation (2.24) using the binomial theorem, we have:

$$T = m_0 c^2 \left\{ 1 + \frac{1}{2} \frac{u^2}{c^2} + \frac{3}{8} \frac{u^4}{c^4} + \; \ldots \; \right\} - m_0 c^2$$

$$= \frac{1}{2} m_0 u^2 + \frac{3}{8} m_0 u^2 \left(\frac{u^2}{c^2} \right) + \; \ldots \; .$$

If $u \ll c$, $u^2 \ll c^2$ and $u^2/c^2 \ll 1$, then

$$T \simeq \tfrac{1}{2} m_0 u^2. \qquad (2.25)$$

Hence, when the speed of the particle is very much less than the speed of light, equation (2.24) for the kinetic energy of a particle approximates very closely to the expression for kinetic energy derived using Newtonian mechanics; but at high speeds comparable to c, Newtonian mechanics is completely inadequate. Notice if u tends to c the kinetic energy tends to infinity, and one would have to do an infinite amount of work to accelerate a particle of finite rest mass up to the speed of light. This illustrates how the speed of light is the limiting speed for particles of finite rest masses.

Using equation (2.10), equation (2.24) can be rewritten:

$$T = mc^2 - m_0 c^2 = (m - m_0)c^2. \qquad (2.26)$$

Equation (2.26) shows that associated with a change in the kinetic energy of the particle there is a change in its mass. Rearranging equation (2.26), we have:

$$mc^2 = T + m_0c^2. \qquad (2.27)$$

It is convenient to introduce a quantity E defined by the relation:

$$E = T + m_0c^2 = mc^2. \qquad (2.28)$$

The quantity E is called the *total energy* of the particle (or sometimes just the energy of the particle). It is equal to the sum of the kinetic energy T and the quantity m_0c^2, which is generally called the rest mass energy of a free particle. (The interconvertibility of rest mass energy and other forms of energy will be illustrated in § 2.7.)

Using equation (2.13), we have:

$$E = mc^2 = \frac{m_0c^2}{\sqrt{(1 - u^2/c^2)}} = m_0c^2 \sec \theta. \qquad (2.29)$$

Squaring,

$$E^2 = m_0{}^2c^4 \sec^2 \theta = m_0{}^2c^4 (\tan^2 \theta + 1).$$

Using equation (2.16) for the momentum of the particle,

$$E^2 = c^2p^2 + m_0{}^2c^4. \qquad (2.30)$$

Equation (2.30) is a very important relation. It relates the total energy E and the momentum p of a particle. Using equation (2.28):

$$T = E - m_0c^2 = \sqrt{(c^2p^2 + m_0{}^2c^4)} - m_0c^2. \qquad (2.31)$$

Equation (2.31) relates p, the momentum of a particle, to its kinetic energy T.

2.4 *Units*

The equations developed in § 2.3 can be applied in any coherent set of units. In S.I. units the speed of light $c = 2\cdot997\,93 \times 10^8$ metre per second. In calculations we shall generally use the approximate value of $3\cdot00 \times 10^8$ metre per second (or m s^{-1}).

The equations valid for high speed particles (i.e. relativistic mechanics) are generally used in atomic and nuclear physics, where it is often convenient to use a system of units based on the electron volt, denoted eV, which is defined as the work done in moving a charge equal to the electronic charge of $1\cdot602 \times 10^{-19}$ coulomb through a potential difference of one volt. Since the work done in moving one coulomb through one volt is one joule (J), the work done in taking $1\cdot602 \times 10^{-19}$ coulomb through one volt is $1\cdot602 \times 10^{-19}$ J. Hence

$$1 \text{ eV} = 1\cdot602 \times 10^{-19} \text{ J} = 1\cdot602 \times 10^{-12} \text{ erg}. \qquad (2.32)$$

19

The following multiples of the electron volt are often used:

$$1 \text{ keV} = 10^3 \text{ eV} = 1 \cdot 602 \times 10^{-16} \text{ J}, \quad (2.33)$$

$$1 \text{ MeV} = 10^6 \text{ eV} = 1 \cdot 602 \times 10^{-13} \text{ J}, \quad (2.34)$$

$$1 \text{ GeV} = 10^9 \text{ eV} = 1 \cdot 602 \times 10^{-10} \text{ J}. \quad (2.35)$$

In the equation

$$E^2 = c^2 p^2 + m_0{}^2 c^4$$

both cp (momentum times velocity of light) and $m_0 c^2$ (rest mass times velocity of light squared) have the dimensions of energy, so that cp and $m_0 c^2$ can be expressed in electron volts. For example, for an electron: $m_0 = 9 \cdot 1083 \times 10^{-31}$ kilogramme so that

$$m_0 c^2 = 9 \cdot 1083 \times 10^{-31} \times (2 \cdot 997 \, 93 \times 10^8)^2$$

$$= 8 \cdot 186 \, 16 \times 10^{-14} \text{ J}$$

$$= 0 \cdot 510 \, 98 \text{ MeV}.$$

Some text-books simply quote the masses of particles in MeV, implying that it stands for $m_0 c^2$.

From equation (2.4) we have:

$$mu = qB\rho,$$

so that

$$p = qB\rho.$$

Hence the momentum of a particle of charge q can be determined from the radius of curvature of its orbit when it is moving perpendicular to a magnetic field of B tesla. If ρ is in metre, and q is Z times the electronic charge:

$$cp = qB\rho c = Z \times 1 \cdot 602 \times 10^{-19} \, B\rho \times 3 \cdot 00 \times 10^8 \text{ J}$$

$$= 300 \, Z \, B\rho \text{ MeV}. \quad (2.36)$$

This is sometimes rewritten as:

$$p = 300 \, Z \, B\rho \text{ MeV}/c. \quad (2.37)$$

A typical value for the magnetic induction of an electromagnet is 1 tesla, which is 10 000 gauss. As an example, we shall calculate the momentum and energy of an electron, the radius of curvature of whose orbit in such a magnetic field is 2 cm = 2×10^{-2} metre. From equation (2.36):

$$cp = 300 \, B\rho = 300 \times 1 \times 2 \times 10^{-2} = 6 \cdot 0 \text{ MeV},$$

$$p = 6 \text{ MeV}/c.$$

20

Since for an electron $m_0c^2 = 0.510\ 98$ MeV, using equation (2.30):

$$E^2 = c^2p^2 + m_0{}^2c^4 = 6^2 + 0.511^2,$$
$$E^2 = 36 + 0.26 = 36.26.$$

Hence

$$E = 6.02 \text{ MeV},$$
$$T = E - m_0c^2 = 6.02 - 0.51 = 5.51 \text{ MeV}.$$

Examples for solution are given at the end of this chapter.

2.5 *Motion in an electric field*

It will be assumed that there is a potential difference V volt between the plates of a large parallel plate capacitor, as shown in fig. 2.4. Let

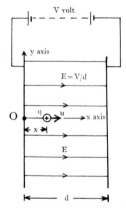

Figure 2.4. The motion of an electric charge in a uniform electric field, for example, in the electric field inside a parallel plate capacitor.

the separation of the plates be d metre. Away from the ends of the capacitor there is a uniform electric field of magnitude $E = V/d$ volt per metre between the plates of the capacitor. Let the x axis of a co-ordinate system be parallel to the direction of the electric field, and let the origin of the co-ordinate system coincide with the positive plate of the capacitor, as shown in fig. 2.4. Let a charge $+q$ coulomb, of rest mass m_0 kilogramme, be released from rest from O, the origin of the co-ordinate system, at the time $t = 0$ such that it accelerates from the positive to the negative plate of the capacitor. For motion in an electric field, the force on the moving charge is qE. When the velocity of the charge is u, from equation (2.18) we have:

$$\frac{d}{dt}\left(\frac{m_0u}{\sqrt{(1 - u^2/c^2)}}\right) = qE.$$

21

Since the *rest* mass m_0 is a constant:

$$\frac{d}{dt}\left(\frac{u}{\sqrt{(1-u^2/c^2)}}\right)=\frac{qE}{m_0}. \tag{2.38}$$

Integrating equation (2.38) with respect to time, if the velocity of the charge after a time t is u, since it starts from rest at $t=0$ we have:

$$\left[\frac{u}{\sqrt{(1-u^2/c^2)}}\right]_0^u=\left[\frac{qE}{m_0}t\right]_0^t,$$

$$\frac{u}{\sqrt{(1-u^2/c^2)}}=\frac{qEt}{m_0}. \tag{2.39}$$

Squaring

$$u^2=(1-u^2/c^2)q^2E^2t^2/m_0{}^2,$$

$$u^2=\frac{q^2E^2t^2}{m_0{}^2(1+q^2E^2t^2/m_0{}^2c^2)},$$

$$u=\frac{qEt}{m_0(1+q^2E^2t^2/m_0{}^2c^2)^{1/2}}. \tag{2.40}$$

Multiplying the top and bottom of the right hand side by m_0c/qEt, we obtain

$$u=\frac{c}{\left\{\dfrac{m_0{}^2c^2}{q^2E^2t^2}+1\right\}^{1/2}}. \tag{2.41}$$

If the time t is very long, that is, if the charge continues to accelerate in the electric field for a long time, the quantity $m_0{}^2c^2/q^2E^2t^2$ becomes small compared with unity, so that as t tends to infinity, u tends to c, though u always remains less than c. Hence the accelerating charge has a limiting speed, which is equal to the speed of light in empty space.

According to Newtonian mechanics, since the electric force acting on the charge is qE, its acceleration a is given by:

$$a=\frac{f}{m_0}=\frac{qE}{m_0}. \tag{2.42}$$

Hence, according to Newtonian mechanics:

$$u=at=\left(\frac{qE}{m_0}\right)t. \tag{2.43}$$

Thus, according to Newtonian mechanics, the velocity u tends to infinity as t tends to infinity and there should be no limiting speed. If $t \ll m_0c/qE$, that is, if the time is not long enough for the charge to

22

gain very much speed, then since $qEt/m_0c \ll 1$ and $q^2E^2t^2/m_0{}^2c^2 \ll 1$, equation (2.40) becomes:

$$u \simeq \frac{qEt}{m_0},$$

in agreement with Newtonian mechanics, which is a satisfactory approximation at low speeds. However, equations (2.40) and (2.43) differ markedly at high speeds.

An experimental determination of the speeds of electrons accelerated by electric fields was carried out by Bertozzi in 1964. The principle of the experiment is outlined in fig. 2.5. Electrons were accelerated

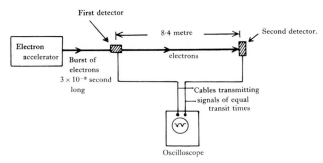

Figure 2.5. Basic principles of the experimental arrangement used by Bertozzi[1] to measure the time of flight of electrons over a distance of 8·4 metre.

by a Van de Graaff generator up to energies of 1·5 MeV. Higher energies were obtained using an electron linear accelerator. The electrons were accelerated in bursts approximately 3×10^{-9} second long. The times of flight of electrons of various kinetic energies were measured over a distance of 8·4 metre. (If the electrons did go at the speed of light the time of flight would be $8·4/3 \times 10^8 = 2·8 \times 10^{-8}$ second, which is long compared with the length of each burst of electrons.) The time of flight was measured by displaying signals from the first and second detectors on an oscilloscope screen, using connecting cables which take the same time to transmit signals from the detectors to the oscilloscope. The separation of the electrical pulses from the two detectors on the oscilloscope screen gave t, the time of flight of the electrons, and their velocity was $8·4/t$. Fuller details of the experiment are given by Bertozzi[1] and Rosser[2 a] The results obtained by Bertozzi are shown in fig. 2.6. From equation (2.28):

$$mc^2 = \frac{m_0c^2}{\sqrt{(1 - u^2/c^2)}} = T + m_0c^2,$$

23

where T is the kinetic energy. Rearranging:

$$\frac{u^2}{c^2} = 1 - \left\{\frac{m_0 c^2}{(T + m_0 c^2)}\right\}^2. \qquad (2.44)$$

Equation (2.44) is plotted in fig. 2.6. The values of kinetic energy shown in fig. 2.6 were calculated by assuming that the kinetic energy was equal to qV, where $q = 1\cdot602 \times 10^{-19}$ coulomb, and V is the potential difference in volts through which the electrons were accelerated. The experimental results of Bertozzi are in excellent agreement with equation (2.44) and the equations valid for high speed particles

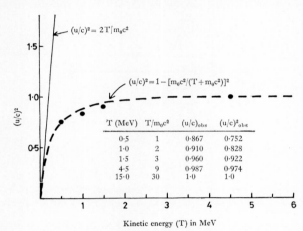

T (MeV)	T/m$_0$c^2	(u/c)$_{obs}$	(u/c)$^2_{obs}$
0·5	1	0·867	0·752
1·0	2	0·910	0·828
1·5	3	0·960	0·922
4·5	9	0·987	0·974
15·0	30	1·0	1·0

Kinetic energy (T) in MeV

Figure 2.6. Experimental results obtained by Bertozzi[1]. The solid line represents the prediction of $(u/c)^2$ according to Newtonian mechanics: $(u/c)^2 = 2T/m_0 c^2$, where $T = \frac{1}{2}m_0 u^2$. The dashed curve is the prediction of special relativity: $(u/c)^2 = 1 - [m_0 c^2/(T + m_0 c^2)]^2$, where m_0 is the rest mass of the electron and $c = 3 \times 10^8$ metre per second is the speed of light in a vacuum. The solid circles are the experimental points. They agree with the theory of high speed particles and show that c is the maximum speed for electrons. (By courtesy of *Am. J. Phys.*)

developed in §2.3. The results show clearly that there is a limiting speed for accelerated electrons, and that the limiting speed is equal to the speed of light in empty space.

Returning to the problem of the charge accelerating between the plates of the capacitor illustrated in fig. 2.4, since the charge is moving along the x axis its velocity is equal to dx/dt. Substituting in equation (2.40) we obtain:

$$\frac{dx}{dt} = \frac{qEt}{m_0\left\{1 + \dfrac{q^2 E^2 t^2}{m_0{}^2 c^2}\right\}^{1/2}}. \qquad (2.45)$$

24

Integrating, assuming that the charge leaves the point $x=0$ at the time $t=0$ and is at the point x at a time t, we have:

$$\int_0^x dx = \int_0^t \frac{qEt\,dt}{m_0\left\{1+\dfrac{q^2E^2t^2}{m_0{}^2c^2}\right\}^{1/2}}. \tag{2.46}$$

Let,

$$1+\frac{q^2E^2t^2}{m_0{}^2c^2} = w.$$

Differentiating:

$$\frac{2q^2E^2t\,dt}{m_0{}^2c^2} = dw$$

or

$$\frac{qEt\,dt}{m_0} = \frac{m_0c^2\,dw}{2qE}.$$

Therefore

$$\int \frac{qEt\,dt}{m_0\{1+q^2E^2t^2/m_0{}^2c^2\}^{1/2}} = \frac{m_0c^2}{2qE}\int \frac{dw}{w^{1/2}} = \frac{m_0c^2}{2qE}\frac{w^{1/2}}{\frac{1}{2}}$$

$$= \frac{m_0c^2}{qE}\left\{1+\frac{q^2E^2t^2}{m_0{}^2c^2}\right\}^{1/2}.$$

Substituting in equation (2.46), we obtain:

$$[x]_0^x = \left[\frac{m_0c^2}{qE}\left(1+\frac{q^2E^2t^2}{m_0{}^2c^2}\right)^{1/2}\right]_0^t,$$

$$x = \frac{m_0c^2}{qE}\left[\left(1+\frac{q^2E^2t^2}{m_0{}^2c^2}\right)^{1/2}-1\right]. \tag{2.47}$$

Equation (2.47) gives the distance x travelled in a time t. Using a little algebra this can be rewritten as:

$$\left(x+\frac{m_0c^2}{qE}\right)^2 - c^2t^2 = \frac{m_0{}^2c^4}{q^2E^2}. \tag{2.48}$$

Equation (2.48) is the equation of a hyperbola, and thus the graph of x against ct is a hyperbola as shown in fig. 2.7, which is drawn for the special case when m_0c^2/qE is equal to unity. Since for an electron m_0c^2 is equal to $0\cdot511$ MeV, for an electron this corresponds to a case where the electric field is $0\cdot511\times10^6$ volt per metre which can be obtained by applying a potential difference of $0\cdot511\times10^6$ volt in the appropriate direction across the plates of a capacitor 1 metre apart.

25

From fig. 2.7 it can be seen that, when x equals 1 metre, ct equals 1·73, so that a negatively charged electron would take a time:

$$ct \div c = 1 \cdot 73/3 \times 10^8 = 5 \cdot 8 \times 10^{-9} \text{ second}$$

to go from the negative to the positive plate in fig. 2.4, when the plates are 1 metre apart and the potential difference between them is $0 \cdot 511 \times 10^6$ volt. Putting $ct = 1 \cdot 73$ and $m_0 c^2/qE = 1$ in equation (2.41):

$$u = \frac{c}{\left\{ \dfrac{m_0{}^2 c^4}{c^2 q^2 E^2 t^2} + 1 \right\}^{1/2}} = \frac{c}{\left\{ \dfrac{1}{1 \cdot 73^2} + 1 \right\}^{1/2}} = 0 \cdot 87c.$$

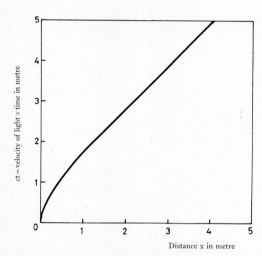

Figure 2.7. A plot of the distance x travelled in a time t by an electric charge accelerating in a uniform electric field.   It is conventional in the theory of special relativity to plot x as abscissa and ct = velocity of light multiplied by time as ordinate.   (In this figure it is assumed that $qE/m_0 c^2$ is equal to unity.)

Hence after travelling 1 metre through a potential difference of $0 \cdot 511 \times 10^6$ volt the speed of the electron is $0 \cdot 87c$.   (This result can also be obtained by applying the law of conservation of energy.   The change of potential energy is $0 \cdot 511$ MeV, so that $T = 0 \cdot 511$ MeV which, for an electron, is in this case numerically equal to $m_0 c^2$. Hence:

$$E = T + m_0 c^2 = 2m_0 c^2 = mc^2 = m_0 c^2/\sqrt{(1 - u^2/c^2)},$$

which gives:

$$1/\sqrt{(1 - u^2/c^2)} = 2; \quad u = \sqrt{3}c/2 = 0 \cdot 866c.)$$

26

According to Newtonian mechanics, the acceleration of the charge in fig. 2.4 is given by equation (2.42), so that:

$$x = \frac{1}{2} at^2 = \frac{1}{2} \left(\frac{qE}{m_0} \right) t^2.$$

If $t \ll m_0 c/qE$, so that the electron does not have enough time to gain very much speed, expanding equation (2.47) using the binomial theorem, we have:

$$x = \frac{m_0 c^2}{qE} \left[1 + \frac{1}{2} \frac{q^2 E^2 t^2}{m_0^2 c^2} + \ldots - 1 \right] \simeq \frac{1}{2} \left(\frac{qE}{m_0} \right) t^2.$$

Even though Newtonian mechanics is completely inadequate at speeds comparable to the speed of light, it is always a very satisfactory approximation at speeds very much less than the speed of light. Even if the speed of an aeroplane is as high as 3000 kilometre per hour, u/c is only $\sim 3 \times 10^{-6}$ and the deviations from Newtonian mechanics are completely negligible. In such cases it is an unnecessary over-elaboration to use the equations of high speed particles (i.e. relativistic mechanics), even though we know that they are a better model of the behaviour of nature than Newtonian mechanics.

2.6 The acceleration of charged particles to high energies

A brief outline of the physical principles of some accelerators will now be given to illustrate the applicability of the laws developed for high speed particles. One simple method of accelerating charged particles is to use a high voltage rectifier circuit. The charged particles are then accelerated by the electric field between the positive and negative terminals. Another method is to use a Van de Graaff generator. In this case also the charges are accelerated by an electric field, and the analysis of § 2.5 is applicable.

Most high energy proton accelerators are developments of the cyclotron, which is shown in schematic form in fig. 2.8. The cyclotron consists of two flat hollow conducting semicircular boxes, generally called dees, placed in a vacuum, as shown in fig. 2.8. The two dees are placed between the poles of an electromagnet, which gives a magnetic field perpendicular to the planes of the dees. Protons are injected from the source S, which is at the centre between the dees, as shown in fig. 2.8. A potential difference is applied between the dees, such that there is an electric field in the gap between the dees. This electric field accelerates the protons towards one of the dees. As the dees are hollow conductors, the electric field inside them is negligible. (Cf. Faraday's ice pail and Faraday's cage experiments.) Hence, once the protons are inside the dee, their velocities remain constant, and

27

they move in circular orbits in the magnetic field, which penetrates into the dees. From equation (2.5) the radius of the orbit is:

$$\rho = \frac{mu}{qB}, \qquad m = m_0/\sqrt{(1-\beta^2)} \qquad (2.49)$$

where m is the relativistic mass, q the charge and u the velocity of the proton in the magnetic induction field B. After going through a semicircle the protons reach the gap between the dees again, as shown in fig. 2.8. According to equation (2.6) the time to go through a semicircle is $\pi\rho/u$, that is, $\pi m/qB$ second. If by this time the potential difference between the dees is reversed in direction, the electric field

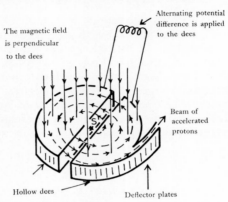

The magnetic field is perpendicular to the dees

Alternating potential difference is applied to the dees

Beam of accelerated protons

Hollow dees

Deflector plates

Figure 2.8. A simplified diagram of a cyclotron; S is the proton source. The path of a typical proton is shown dotted.

is in the opposite direction to previously, and the protons are accelerated by the electric field and gain energy again as they cross the gap between the two dees. Due to their acceleration in the electric field between the dees, u, the velocity of the protons, is increased so that, according to equation (2.49), they move in a semicircle of larger radius before reaching the gap again, as shown in fig. 2.8. The time taken is $\pi m/qB$, which is the same as the time for the previous semicircle, provided the mass m can be treated as a constant. Hence if the potential difference applied to the dees changes direction every $\pi m/qB$ second, the protons are accelerated every time they cross the gap between the two dees. The reversal of potential difference can be obtained by applying an alternating potential difference to the dees. If the potential difference is to reverse in direction every $\pi m/qB$ second the period of the alternating potential difference must be $2\pi m/qB$, so that the frequency n of the potential difference must be:

$$n = \frac{qB}{2\pi m} = \frac{qB\sqrt{(1-u^2/c^2)}}{2\pi m_0}. \qquad (2.50)$$

28

Provided the variation of mass with velocity can be neglected, the protons can be accelerated by applying a potential difference of fixed frequency to the dees. As the energy of the proton, and hence its velocity, increases, the radius of its orbit is increased. Since from equation (2.4)

$$p = mu = qB\rho$$

for a fixed value of magnetic field, the maximum momentum that can be obtained from the cyclotron is proportional to ρ the radius of the dees. For a discussion of typical numerical values see Problem (2.9). The cyclotron has been used to accelerate protons to about 40 MeV. (Electrons can only be accelerated up to about 50 keV in a cyclotron.) At higher energies the variation of mass with velocity becomes important, and as u, and hence m, increases, according to equation (2.6), the time taken to go in a semicircle increases so that, if n and B were fixed, after every half cycle the protons would tend to get further and further behind the maximum potential difference across the dees. This loss of resonance can be overcome in two ways. From equation (2.50), as u increases, the frequency necessary for resonance goes down. In the 184-inch frequency-modulated cyclotron at the University of California they reduced the frequency of the potential difference applied to the dees continuously after injecting the protons. By reducing the frequency from $22 \cdot 9$ MHz at injection to $15 \cdot 8$ MHz when the protons reached the outside of the dees, they were able to accelerate protons up to 350 MeV. The other method of overcoming resonance is to increase the strength of the magnetic field in such a way that as m increases B/m remains constant. One can then keep the frequency n constant. This is the principle of the synchrotron, which has been used to accelerate electrons up to energies > 400 MeV. In proton synchrotrons both n and B are varied in such a way that not only is equation (2.50) satisfied, but in equation (2.49) the radius of the orbit remains constant. This means that instead of dees one can use doughnut-shaped vacuum chambers, and the magnetic field need only extend over the dimensions of the doughnut-shaped vacuum chamber. The NIMROD proton synchrotron of the Rutherford High Energy Laboratory (shown on the front cover) can accelerate protons up to 7 GeV. The CERN proton synchrotron can accelerate protons up to an energy of 25 GeV, and a bigger machine which can accelerate protons up to 300 GeV is planned.

The successful design of high energy accelerators shows that the theory developed in § 2.3, where it was assumed that the mass varies with velocity, but the total electric charge on a particle is constant, is satisfactory up to very high energies.

2.7 *The equivalence of mass and energy*
From equation (2.26):

$$T = (m - m_0)c^2. \tag{2.51}$$

29

If the kinetic energy T increases to $T + \Delta T$, the mass m must increase to $m + \Delta m$ such that

$$T + \Delta T = (m + \Delta m - m_0)c^2. \qquad (2.52)$$

Subtracting equation (2.51) from equation (2.52):

$$\Delta T = \Delta m c^2 \qquad (2.53)$$

or

$$\Delta m = \Delta T / c^2. \qquad (2.54)$$

Equation (2.54) shows that as the kinetic energy of a particle increases by ΔT the mass of the particle increases by $\Delta T/c^2$. If we know the increase in the kinetic energy of a particle, we can calculate the change in its mass and vice versa. On the basis of relativistic arguments, Einstein suggested that mass (or inertia) must be associated with *all* forms of energy, the relation between the energy E of a system and its mass m being:

$$E = mc^2. \qquad (2.55)$$

If the total mass of a system is known, its total energy can be calculated. Energy and mass are measures of different properties of matter. For example, the energy of a particle is a measure of its capacity to do work, whereas its mass is a measure of the resistance of the particle to changes in its motion. According to the law of the equivalence of mass and energy, the changes in energy and mass are proportional to each other, and related by equation (2.55). Our approach, in this section, will be to postulate equation (2.55) as a plausible generalization of equation (2.53), and then to show that predictions based on equation (2.55) are in agreement with experiment.

(a) Nuclear reactions

The most familiar application of equation (2.55) is in nuclear reactions. For example, the mass of a helium nucleus is $6 \cdot 645\ 64 \times 10^{-27}$ kg whilst the mass of two free protons $(2 \times 1 \cdot 672\ 39 \times 10^{-27}$ kg) plus the mass of two free neutrons $(2 \times 1 \cdot 674\ 70 \times 10^{-27}$ kg) is $6 \cdot 694\ 18 \times 10^{-27}$ kg, which is $0 \cdot 048\ 54 \times 10^{-27}$ kg greater than the mass of a helium nucleus. In the nuclear reaction

$$2{}_1^1\text{H} + 2{}_0^1\text{n} \rightarrow {}_2^4\text{He} \qquad (2.56)$$

where ${}_1^1\text{H}$ stands for a proton, ${}_0^1\text{n}$ for a neutron and ${}_2^4\text{He}$ for a helium nucleus, the total rest mass on the right-hand side is less than the total rest mass on the left-hand side by $0 \cdot 048\ 54 \times 10^{-27}$ kg. According to equation (2.55), the energy to be associated with the rest masses is less on the right-hand side by:

$$\Delta E = (\Delta m)c^2 = 0 \cdot 048\ 54 \times 10^{-27} \times (3 \cdot 00 \times 10^8)^2\ \text{J}.$$

Using equation (2.34) we get:

$$\Delta E = 27\ \text{MeV}.$$

If the *total* energy is conserved in the reaction given by equation (2.56), an amount of kinetic energy Q equal to 27 MeV must be liberated in the reaction given by equation (2.56). This illustrates how, using equation (2.55), the kinetic energy released in a nuclear reaction can be calculated from the total changes in the rest masses of the particles. For example, in the reaction

$$A + B \rightarrow C + D + Q$$

the increase in kinetic energy Q is given by:

$$Q = (m_A + m_B - m_C - m_D)c^2. \tag{2.57}$$

The masses of the particles can be determined accurately using mass spectrographs. Equation (2.55) has been shown to be in agreement with experiment in all nuclear reactions.

The nuclear reaction given by equation (2.56) cannot be performed in practice as one cannot make two protons and two neutrons coincide at the same point of space at the same time. The type of nuclear reaction in which particles join together to form heavier nuclei is called fusion, or a thermonuclear reaction. It is believed that thermonuclear reactions are the main source of the energy liberated in stars, such as the sun. It is believed that there are two main cycles of nuclear reactions in stars, namely the proton–proton chain, which predominates in small stars less massive than the sun, and the carbon–nitrogen–oxygen cycle which predominates in larger stars. The net result of the series of nuclear reactions, in both cases, is to convert four protons into a helium nucleus plus two positive electrons (positrons), two neutrinos, gamma rays and about 26 MeV of thermal energy per helium nucleus formed.

The energy reaching the top of the earth's atmosphere from the sun does so at the rate of $1 \cdot 35 \times 10^3$ watt per square metre (W m^{-2}). The distance from the earth to the sun is $1 \cdot 5 \times 10^{11}$ metre. The total energy radiated by the sun per second should be equal to the energy crossing a sphere of radius $1 \cdot 5 \times 10^{11}$ metre at the rate of $1 \cdot 35 \times 10^3$ W m^{-2}. Hence the total energy radiated by the sun per second is:

$$\Delta E = 4\pi(1 \cdot 5 \times 10^{11})^2 \times 1 \cdot 35 \times 10^3 \sim 4 \times 10^{26}\ \text{J}$$

Using equation (2.55), the total mass lost by the sun per second is:

$$\Delta m = \frac{\Delta E}{c^2} = \frac{4 \times 10^{26}}{9 \times 10^{16}} \sim 4 \cdot 4 \times 10^9\ \text{kg}\ (\sim 4 \cdot 4 \times 10^6\ \text{ton}).$$

Thus the mass of the sun decreases by over 4×10^9 kilogramme per second. This is negligible, however, compared with the total mass of the sun, which is $1 \cdot 98 \times 10^{30}$ kg.

Another type of nuclear reaction is fission. As the sizes of atomic nuclei increase, the protons and neutrons in the nuclei are not as tightly bound as in smaller nuclei. If a large atomic nucleus, such as uranium, splits into two approximately equal halves, the protons and neutrons in the separate halves are more tightly bound together than they were in the original nucleus. The total mass decreases and kinetic energy is liberated, when fission takes place. A typical example of the fission of a ^{235}U nucleus, following the capture of a neutron (denoted ^1_0n), is

$$^{235}_{92}\text{U} + ^1_0\text{n} \rightarrow ^{236}_{92}\text{U} \rightarrow ^{140}_{54}\text{Xe} + ^{94}_{38}\text{Sr} + 2^1_0\text{n} + \gamma + 200 \text{ MeV}.$$

(b) Electromagnetic radiation

Electromagnetic waves carry energy. If inertia must be associated with all forms of energy, electromagnetic waves must carry momentum and so be capable of exerting a pressure. The existence of the radiation pressure due to light was shown by Lebedew in 1900.

Electromagnetic radiation is emitted in the form of discrete quanta of energy called photons. According to Planck, if the frequency of the radiation is ν, the energy of each quantum, or photon, is $E = h\nu$, where $h = 6 \cdot 625 \times 10^{-34}$ joule second is Planck's constant. If the energy of the quantum or photon is in the MeV range, it is generally referred to as a γ-ray. Einstein used the photon concept to interpret the photoelectric effect. Compton used the idea of individual photons of energy $h\nu$ carrying momentum $h\nu/c$ to interpret the scattering of X-rays. The existence of individual photons travelling at the speed of light is now a generally accepted experimental fact. For a particle of rest mass m_0, we have from equations (2.10), (2.14) and (2.29):

$$m = \frac{m_0}{\sqrt{(1 - u^2/c^2)}}; \quad p = \frac{m_0 u}{\sqrt{(1 - u^2/c^2)}}; \quad E = \frac{m_0 c^2}{\sqrt{(1 - u^2/c^2)}}.$$

As u tends to c, the denominator tends to zero in each case. If the rest mass m_0 also tends to zero, then each of the above quantities can remain finite. If $m_0 \rightarrow 0$ when $u \rightarrow c$, such that $m_0/(1 - u^2/c^2)^{1/2}$ equals k, then

$$m = k; \quad p = kc; \quad E = kc^2.$$

From Planck's relation, for a photon E equals $h\nu$, so that k must equal $h\nu/c^2$. Hence a photon of energy $h\nu$ should have a linear momentum $h\nu/c$ and an inertial mass $h\nu/c^2$ associated with it.

(c) Pair production

Soon after the discovery of the positive electron (or positron) in 1932, pair production was discovered. If a photon has an energy

32

$hv > 2m_0c^2$, where m_0 is the rest mass of the electron, then the photon can give rise to an electron–positron pair, that is, create a positive and a negative electron in the electric field of an atomic nucleus. In pair production, radiant energy in the form of a photon is transformed into the rest mass energies of a positron and an electron. If the energy of the photon exceeds $2m_0c^2$ the excess energy is distributed as kinetic energy between the electron, the positron and the atomic nucleus. The presence of the atomic nucleus is necessary, if both energy and momentum are to be conserved in the process.

When a positron slows down it can be attracted to a negative electron and the two annihilate each other, giving rise to either two or three γ-rays. Positron annihilation is an example of the conversion of rest mass energy into radiant energy.

(d) *Meson production*

It is now believed that the nuclear forces holding the protons and neutrons in atomic nuclei together are due to a meson field, which involves very short-lived particles called mesons. When fast protons or neutrons collide with atomic nuclei, positively charged, negatively charged and neutral π-mesons can be produced, provided the kinetic energy of the incident particle is high enough (cf. § 5.5(b)). The charged π-mesons have electric charges of magnitude $1 \cdot 602 \times 10^{-19}$ coulomb, and rest masses of $139 \cdot 6 \, \text{MeV}/c^2$ (or $273 \cdot 2$ times the rest mass of an electron). Neutral π-mesons have a rest mass of $135 \cdot 0$ MeV/c^2 (or $246 \cdot 4$ times the rest mass of an electron). When charged π-mesons are at rest, they live for an average time of $2 \cdot 55 \times 10^{-8}$ second, before decaying into a μ-meson and a neutrino. Examples of π-meson decay are shown in Plate 1. The neutrino is a neutral particle of zero rest mass. The μ-meson has a charge of $\pm 1 \cdot 602 \times 10^{-19}$ coulomb, a mass of $105 \cdot 7 \, \text{MeV}/c^2$ (or $206 \cdot 8$ electron masses) and a mean lifetime (at rest) of $2 \cdot 2 \times 10^{-6}$ second. The μ-meson decays into an electron and two neutrinos. Examples of μ-meson decay following π-meson decay are shown in Plate 1. The neutral π-meson (or π^0-meson) has no electric charge, its average lifetime, when it is at rest, is $\sim 10^{-16}$ second and it decays into two γ-rays. When the protons and neutrons striking atomic nuclei have sufficient kinetic energy, they can also produce K-mesons (of rest mass $494 \, \text{MeV}/c^2$), protons, negative protons, hyperons, resonance particles, etc. In these processes some of the kinetic energy of the incident particle is converted into the rest mass energies of the created particles. Mesons can also be produced when high energy γ-rays (photons) collide with atomic nuclei. In this case radiant energy is converted into rest mass energy.

(e) *Particle decays*

Let a particle of rest mass M decay at *rest* into a particle of rest mass m_1, having momentum p_1 and total energy E_1 and into a particle

33

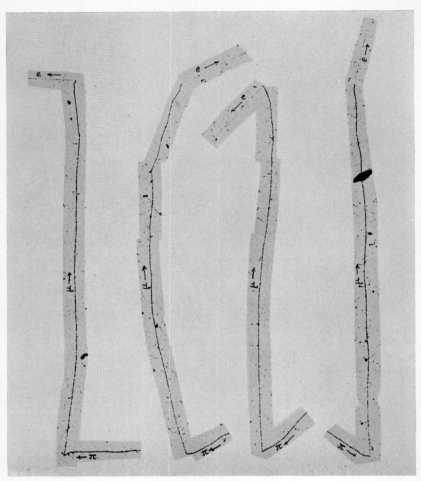

PLATE 1

When charged particles pass through a photographic emulsion, they produce
ionization in some of the silver bromide grains through which they pass.
When the photographic plate is developed these grains develop black
just as if light had struck them. Charged particles therefore give rise to
a series of black grains, that is give a visible track on the photographic
plate. Neutral particles, such as neutrinos, produce no ionization and
give no visible track on the photographic plate. Four examples of the
decay of a π-meson are shown. In each case a charged π-meson comes
to rest in the photographic emulsion, and decays into a charged μ-meson
and a neutrino, which produces no visible track. The μ-mesons all
have a range corresponding to a kinetic energy of 4·1 MeV. Each
μ-meson decays into an electron (visible track) plus two neutrinos.
(Reproduced by permission of Professor C. F. Powell and of the Physical
Society.)

34

of rest mass m_2, having momentum p_2 and total energy E_2 as shown in fig. 2.9. If linear momentum is to be conserved m_1 and m_2 must go off in opposite directions, such that

$$p_1 = p_2. \tag{2.58}$$

The law of conservation of energy gives:

$$E_1 + E_2 = Mc^2. \tag{2.59}$$

From equation (2.58), multiplying both sides by c and squaring we have:

$$c^2 p_1{}^2 = c^2 p_2{}^2.$$

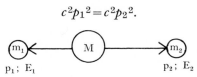

Particle M is at rest when it decays

Figure 2.9. A particle of rest mass M decays at rest into two particles of rest masses m_1 and m_2, which go in opposite directions, so as to conserve linear momentum.

Using equation (2.30), $c^2 p_1{}^2 = E_1{}^2 - m_1{}^2 c^4$ and $c^2 p_2{}^2 = E_2{}^2 - m_2{}^2 c^4$. Hence,

$$E_1{}^2 - m_1{}^2 c^4 = E_2{}^2 - m_2{}^2 c^4$$

or

$$E_1{}^2 - E_2{}^2 = (E_1 + E_2)(E_1 - E_2) = (m_1{}^2 - m_2{}^2)c^4. \tag{2.60}$$

Dividing equation (2.60) by equation (2.59), one obtains:

$$E_1 - E_2 = \frac{(m_1{}^2 - m_2{}^2)c^2}{M}. \tag{2.61}$$

Adding equations (2.61) and (2.59) we have:

$$E_1 = \frac{(M^2 + m_1{}^2 - m_2{}^2)c^2}{2M}. \tag{2.62}$$

Subtracting equation (2.61) from equation (2.59) one obtains:

$$E_2 = \frac{(M^2 + m_2{}^2 - m_1{}^2)c^2}{2M}. \tag{2.63}$$

Using equations (2.30) and (2.58) we find that the momenta p_1 and p_2 of the two particles are given by:

$$cp_1 = cp_2 = \sqrt{(E_1{}^2 - m_1{}^2 c^4)} = \sqrt{(E_2{}^2 - m_2{}^2 c^4)}. \tag{2.64}$$

The theory developed above can be applied to radioactive decay, photon emission, meson decay, etc. For example, a π-meson decays into a μ-meson and a neutrino, as shown in Plate 1. Experiments have shown that $m_\pi c^2 = 139 \cdot 6$ MeV, $m_\mu c^2 = 105 \cdot 7$ MeV and the neutrino has zero rest mass. If the π-meson decays at rest, substituting into equation (2.62), we obtain:

$$E_\mu = \frac{139 \cdot 6^2 + 105 \cdot 7^2 - 0^2}{2 \times 139 \cdot 6} = 109 \cdot 8 \text{ MeV.}$$

Hence the kinetic energy of the μ-meson is:

$$T_\mu = E_\mu - m_\mu c^2 = 109 \cdot 8 - 105 \cdot 7 = 4 \cdot 1 \text{ MeV.}$$

This result is confirmed by experiment. For example, the ranges of the μ-mesons arising from the decay of the stationary π-mesons in Plate 1 are in agreement with this value.

(f) Discussion

In this section it has been illustrated how various forms of energy can be converted into other forms of energy. Using equation (2.55) it was possible to estimate the changes in energy from the changes in mass. It is possible to account for the dynamics of all these events, such as pair production, meson production, meson decay, etc., by assuming that linear momentum is conserved in these events and that the total energy, which includes the rest mass energy, is also conserved in these events. If mass and energy are proportional to each other, the law of conservation of mass is equivalent to the law of conservation of energy and they are often combined into the *law of conservation of mass–energy*. For a fuller discussion the reader is referred to Rosser[(2b)].

2.8 *High energy particles and the principle of relativity*

Generally, the laws of high speed particles are developed using the theory of special relativity, e.g. see Rosser[(2b)]. For this reason the theory of high speed particles is generally called relativistic mechanics. The conventional approach via special relativity tends to give the illusion that special relativity and the Lorentz transformations are pre-requisites, before the laws of high speed particles can be developed. This is not correct. In this chapter the laws of high speed particles were developed from the results of experiments in the laboratory. This approach shows that modern *experiments* show that Newtonian mechanics is completely inadequate at high speeds, and one must be prepared to re-interpret all concepts based on Newtonian mechanics, including the Galilean transformations. In this section it will be *indicated* how, if the new *experimental* laws developed for high speed particles are to obey the principle of relativity, then the co-ordinates and time must be transformed, using the Lorentz transformations and not the Galilean transformations. The full development and inter-pretation of the Lorentz transformations will be deferred until Chapter 4, when the conventional approach to special relativity, based on the principle of the constancy of the speed of light, will be developed. The arguments in the rest of this section may appear a little concise and subtle at a first reading. The reader can return to this section after reading Chapter 4. All we are trying to indicate in this section is that the Lorentz transformations follow, if it is assumed that the new *experimental* laws developed for high speed particles in this chapter, obey the principle of relativity, that is, hold in all inertial reference frames. *If he is ready to believe this, the reader can move on to Chapter 3.*

Consider an astronaut inside a spaceship coasting along with uniform velocity relative to the fixed stars. The astronaut is an inertial observer. If the astronaut carried out experiments on high

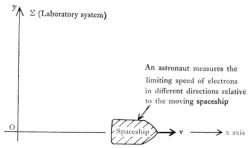

Figure 2.10. The limiting speed of accelerated electrons is measured inside the moving spaceship. If space is isotropic, the limiting speed should be the same in all directions relative to the spaceship.

speed particles, it seems reasonable to assume that on the basis of such experiments the astronaut would develop the same experimental laws for high speed particles, as we developed in this chapter on the basis of terrestrial experiments. For example, the astronaut should find that the mass of a particle varied with its velocity relative to the spaceship. In the theory of special relativity it is *assumed* that all the laws of physics, including those of high speed particles obey the principle of relativity. Some of the consequences of extending the principle of relativity to the experimental laws of high speed particles will now be discussed.

Consider a spaceship moving with uniform velocity v along the x axis relative to the laboratory system Σ, as shown in fig. 2.10. The symbol Σ is the Greek capital letter sigma. If the laws developed for high speed particles obey the principle of relativity, and if the astronaut on the spaceship repeated Bertozzi's experiment, which was described in § 2.5, then the astronaut should find that the accelerated electrons had a limiting speed relative to his spaceship. If space is isotropic, that is, if all directions in space are equivalent, the astronaut should find experimentally that the limiting speed of accelerated electrons was the same in all directions, relative to the spaceship.

A plausibility argument will now be given to show that the limiting speed of accelerated electrons should have the same numerical value relative to the spaceship as relative to the earth, provided the same units of length and time are used in both reference frames. For purposes of discussion only it will be assumed initially that the numerical value of the limiting speed of accelerated electrons, measured relative to the spaceship in fig. 2.10, depends on the speed of the spaceship relative to the earth. The astronaut can transmit a radio message to his base station on the earth giving the measured value of the limiting speed of accelerated electrons relative to his spaceship. An observer on the earth or the astronaut can measure v, the speed of the spaceship relative to the earth, for example, by using radar methods. The experiment can be repeated for various values of v and a table made, relating the limiting speeds of electrons, measured relative to the spaceship, to the speed of the spaceship relative to the earth. This information can be sent by radio to the astronaut. Thereafter, by measuring the speed of accelerated electrons relative to his spaceship and using the table, the astronaut could determine the speed of his spaceship relative to the earth, without looking at anything external to the spaceship. This is contrary to the assumption that the laws of physics are the same in all inertial reference frames, and that they do not contain the speed of the astronaut relative to any absolute reference frame. The way out of this dilemma is to conclude that the numerical value of the limiting speed of accelerated electrons relative to the spaceship is the same, and equal to the terrestrial value of $3 \cdot 0 \times 10^8$ metre per second, whatever the speed of the spaceship relative to the earth. This principle will be called the *principle of the constancy of the limiting speed of particles*, according to which the numerical value of the limiting speed of accelerated particles is the same in all directions in all inertial reference frames and equal to $c = 3 \cdot 0 \times 10^8$ metre per second. This principle will now be used to develop the Lorentz transformations.

Consider two inertial reference frames Σ and Σ', such as two spaceships passing each other in outer space, with Σ' moving with a high uniform velocity v along the x axis relative to Σ, as shown in fig. 2.11. Let the origins coincide at times $t = t' = 0$, as shown in fig. 2.11 a. Let an electron pass the origins of Σ and Σ' at this instant, and let it move along the $+x'$ axis of Σ' with a speed very, very close to the limiting speed c relative to Σ'. Let the electron be detected at an event at P at x' at a time t' relative to Σ', as shown in fig. 2.11 b. Since the electron moves with a speed very close to c relative to Σ', it moves a distance very nearly equal to ct' in a time t' relative to Σ', so that relative to Σ':

$$x' \simeq ct'. \tag{2.65}$$

Now, Σ' moves with uniform velocity v along the x axis relative to Σ, so that one would expect the speed of the electron, which is moving along the $+x$ axis, to be greater relative to Σ than Σ', but still less

38

than the limiting value c. Hence the speed of the electron must be very, very close to the limiting value c relative to Σ also. If the electron is detected at the event at P at x at a time t relative to Σ by the detector, which may be moving relative to both Σ and Σ', then relative to Σ we should have:

$$x \simeq ct. \tag{2.66}$$

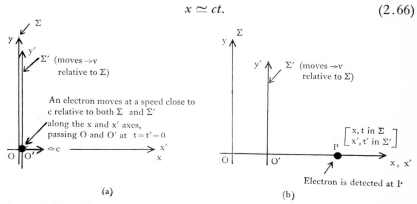

(a)

(b)

Figure 2.11. (a) An electron travelling at a speed very close to c along the common x axis relative to both Σ and Σ' passes the origins O and O' at $t=t'=0$. (b) The electron is detected at P, which is measured to be at x at a time t relative to Σ and at x' at a time t' relative to Σ'.

Since in fig. 2.11 the origin of Σ' moves relative to the origin of Σ during the time the electron moves from the origins of Σ and Σ' until it is detected at P, then $x' < x$. Hence t' must be less than t if equations (2.65) and (2.66) are both to be satisfied. Thus time cannot be absolute, as it was assumed to be in Newtonian mechanics. Thus the concept of absolute time which leads to the Galilean transformations must be abandoned, if the experimental laws of high speed particles are to obey the principle of relativity.

Let the electron passing the origins of Σ and Σ' at $t=t'=0$, not now move along the x axis of Σ, but let it travel at a speed very, very close to c relative to both Σ and Σ' as shown in fig. 2.12 a. Let the electron be detected at an event which has co-ordinates x', y', z' at a time t' relative to Σ' and co-ordinates x, y, z at a time t relative to Σ, as shown in fig. 2.12 b. In a time t' the electron travels a distance $\sqrt{(x'^2+y'^2+z'^2)}$ at a speed very close to c relative to Σ'. Hence:

$$\sqrt{(x'^2+y'^2+z'^2)} \simeq ct'$$

or

$$x'^2+y'^2+z'^2-c^2t'^2 \simeq 0. \tag{2.67}$$

Similarly, relative to Σ:

$$\sqrt{(x^2+y^2+z^2)} \simeq ct,$$
$$x^2+y^2+z^2-c^2t^2 \simeq 0. \tag{2.68}$$

39

Notice the same value c was used for the limiting speed of the electron in both equation (2.67) and equation (2.68). The appropriate transformations which transform equation (2.67) into equation (2.68) are the Lorentz transformations:

$$x' = \frac{(x-vt)}{\sqrt{(1-v^2/c^2)}} = \gamma(x-vt), \tag{2.69}$$

$$y' = y, \tag{2.70}$$

$$z' = z, \tag{2.71}$$

$$t' = \frac{(t-vx/c^2)}{\sqrt{(1-v^2/c^2)}} = \gamma(t-vx/c^2), \tag{2.72}$$

where v is the velocity of Σ' relative to Σ, and

$$\gamma = \frac{1}{\sqrt{(1-v^2/c^2)}}. \tag{2.73}$$

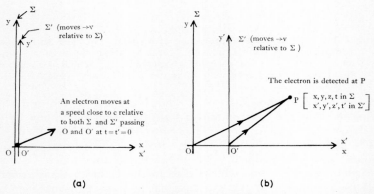

(a)   (b)

Figure 2.12.   (a) An electron travelling at a speed very close to the limiting speed c relative to both Σ and Σ' passes O and O' at $t=t'=0$.   (b) The electron appears to go along the path OP relative to Σ and along the path O'P relative to Σ'.

For example:

$$x'^2 + y'^2 + z'^2 - c^2t'^2 = \gamma^2(x-vt)^2 + y^2 + z^2 - c^2\gamma^2(t-vx/c^2)^2$$

$$= \gamma^2(x^2 - 2xvt + v^2t^2 - c^2t^2 + 2vxt - v^2x^2/c^2)$$

$$+ y^2 + z^2$$

$$= \gamma^2x^2(1-v^2/c^2) + y^2 + z^2 - c^2\gamma^2t^2(1-v^2/c^2)$$

$$= x^2 + y^2 + z^2 - c^2t^2.$$

Hence the Lorentz transformations transform equation (2.67) into equation (2.68). The full development of the Lorentz transformations

40

and their physical interpretation will be deferred until Chapter 4. In this section we have merely tried to indicate how they follow, if it is assumed that the experimental laws developed for high speed particles obey the principle of relativity.

Returning to equations (2.65) and (2.66), it will be assumed that equation (2.66) is valid, that is $x = ct$. Using the Lorentz transformatons and $x = ct$, we have:

$$\frac{x'}{t'} = \frac{\gamma(x - vt)}{\gamma(t - vx/c^2)} = \frac{(ct - vt)}{(t - vct/c^2)} = \frac{c(ct - vt)}{(ct - vt)} = c.$$

Hence, if the co-ordinates and time are transformed using the Lorentz transformations, it is possible to satisfy both $x = ct$ and $x' = ct'$ in fig. 2.11. In addition to different measures of the distances x and x' measured relative to Σ and Σ' respectively the times t and t' must also be different such that equations (2.65) and (2.66) can be satisfied. It will be illustrated in Chapter 4 how this new concept of time is in accord with experiment.

Historically the theory of special relativity evolved from classical optics and electricity and magnetism. An account of the historical development of special relativity is given in Chapter 3. This was preceded by the account of high speed particles in this chapter so as to convince the reader that Newtonian mechanics, and concepts based on Newtonian mechanics, such as the Galilean transformations, are completely inadequate at high speeds. There is generally a natural bias in students' minds, based on their everyday experiences, in favour of the concepts of Newtonian mechanics, which makes the necessity for the theory of special relativity difficult to appreciate when special relativity is approached in the traditional way via optics and electromagnetism, where the experimental deviations from the Galilean transformations are generally very, very small. With historical hindsight, the reader can approach Chapter 3 and the theory of special relativity, already familiar with the inadequacies of Newtonian mechanics and the Galilean transformations at high speeds.

References
1. BERTOZZI, W., *Am. J. Phys.*, **32** (1964), 551.
2. ROSSER, W. G. V., *Introductory Relativity* (Butterworths, London, 1967). (a) p. 129; (b) Ch. 5.

Problems

Velocity of light $c = 3 \cdot 00 \times 10^8$ m s^{-1}.
Electronic charge $e = 1 \cdot 602\ 06 \times 10^{-19}$ C.
Electron rest mass $m = 9 \cdot 1083 \times 10^{-31}$ kg.
$\equiv 0 \cdot 510\ 976$ MeV/c^2.
Proton rest mass $= 1 \cdot 672\ 39 \times 10^{-27}$ kg.
$\equiv 938 \cdot 211$ MeV/c^2.

41

2.1. Find the ratio of the relativistic mass m to the rest mass m_0 for particles of speeds (a) 300 kilometre per hour, (b) 3000 kilometre per hour, (c) $0 \cdot 1c$, (d) $0 \cdot 5c$, (e) $0 \cdot 9c$, (f) $0 \cdot 99c$, (g) $0 \cdot 999c$, (h) $0 \cdot 9999c$.

2.2. Calculate the velocities of electrons accelerated through potential differences of (a) 10 000; (b) 100 000; and (c) 1 000 000 V. What is the ratio of the relativistic mass to the rest mass in each instance? (Hint: The kinetic energy of the electron is equal to charge times potential difference, then use equation (2.24).)

2.3. Calculate the velocities of protons of kinetic energies (a) 10; (b) 100; and (c) 1000 MeV. (Hint: Use equation (2.24).)

2.4. Calculate the radii of curvature of electrons of velocities (a) $0 \cdot 3c$; (b) $0 \cdot 8c$; (c) $0 \cdot 99c$; and (d) $0 \cdot 999c$ in a magnetic field of strength 1 tesla (or 10 000 gauss). (Hint: use equations (2.16)) and (2.37).

2.5. Calculate the amount of work in MeV that must be done to increase the velocity of an electron (a) to half the velocity of light; (b) to three quarters the velocity of light. What is the ratio of the relativistic mass to the rest mass in each instance?

2.6. Through what potential difference must an electron fall if, according to Newtonian mechanics, its velocity is to equal the velocity of light? What speed does the electron actually acquire according to the theory of high speed particles.

2.7. Show that the rest mass of a particle is given by:

$$m_0 = \frac{p^2 c^2 - T^2}{2Tc^2},$$

where p is its momentum and T its kinetic energy. (Hint: Put $E = T + m_0 c^2$ in $E^2 = c^2 p^2 + m_0{}^2 c^4$.) Calculate the rest mass of a particle if its momentum is 130 MeV/c when its kinetic energy is 50 MeV.

2.8. A particle of charge $1 \cdot 60 \times 10^{-19}$ C is accelerated through a potential difference of 5×10^8 V. It is then passed through a uniform magnetic field of strength 1 tesla. The radius of curvature of the orbit of the particle in the magnetic field is $3 \cdot 62$ m, when the plane of the orbit is perpendicular to the magnetic field. Calculate the rest mass and velocity of the particle. (Hint: Calculate the kinetic energy and the momentum of the particle, then use the method of Problem 2.7.)

2.9. In a cyclotron the strength of the magnetic induction is $1 \cdot 5$ tesla. Show, using equation (2.50), that the frequency for resonance is equal to 23 MHz. Using equation (2.37), show that if the radius of the dees is $0 \cdot 4$ m the protons have momenta of 180 MeV/c when they reach the outside of the dees. Show that their velocity is $5 \cdot 6 \times 10^7$ m s^{-1}. Use equation (2.24) to show that this corresponds to a kinetic energy of 17 MeV.

42

CHAPTER 3
historical development of the theory of special relativity

3.1 *Introduction*

NEWTONIAN mechanics was developed mainly in the seventeenth century, whereas the wave theory of physical optics and the theory of classical electromagnetism were developed mainly in the nineteenth century. For example, Young re-introduced the wave theory to interpret interference experiments in 1801, and Oersted discovered the magnetic effect due to an electric current in 1820. The enormous successes of Newtonian mechanics naturally led people to assume that the Galilean transformations were correct. After reading Chapter 2 the reader will realize that twentieth-century experiments have shown that Newtonian mechanics is inadequate at high speeds and by now the reader is presumably prepared to accept that the Galilean transformations must therefore be replaced. However, if we try to cast our minds back to the nineteenth century, at that time no experiments on very high speed particles had been performed, and for centuries Newton's theories had appeared to be the perfect examples for all physical theories. Thus it was natural in the nineteenth century to assume that the Galilean transformations, which were based on Newtonian mechanics, were correct, and could be applied to the laws of optics and of electricity and magnetism. It will now be illustrated how this assumption led to difficulties which could only be resolved by abandoning the Galilean transformations.

3.2 *Optical experiments and the principle of relativity*

Consider an *absolute* inertial reference frame Σ in which the speed of light in empty space is the same, and equal to c, in all directions of empty space, as shown in fig. 3.1 *a*. Consider a reference frame Σ' moving with uniform velocity v relative to the absolute system Σ along their common x axis, as shown in fig. 3.1 *b*. The Galilean velocity transformations will be used to calculate the speed of light in various directions of empty space relative to the reference frame Σ', which is moving relative to the absolute system Σ. For light moving in the $+x$ direction of Σ, $u_x = c$; $u_y = 0$. Applying the Galilean velocity transformations, equations (1.18) and (1.19), we have relative to Σ:

$$u_x' = u_x - v = c - v; \quad u_y' = 0,$$

as illustrated in fig. 3.1 *b*. For light moving in the negative x direction relative to Σ, $u_x = -c$; $u_y = 0$. Using the Galilean velocity transformations, equations (1.18) and (1.19), relative to Σ', we have:

$$u_x' = u_x - v = -c - v = -(c+v); \quad u_y' = 0,$$

as illustrated in fig. 3.1 b. Light moving parallel to the y' axis of Σ' must have a velocity c having components $u_x = -v$ and $u_y = \sqrt{(c^2 - v^2)}$ in Σ, so that using equations (1.18) and (1.19) we would then have $u_x' = 0$ and $u_y' = \sqrt{(c^2 - v^2)}$, as illustrated in fig. 3.1 b. Thus if it were assumed that the Galilean velocity transformations were correct, the speed of light in empty space should be different in different directions in a reference frame Σ' moving relative to the absolute system Σ, in which the speed of light in empty space was the same in all directions. In the nineteenth century it was argued that, if there were such an

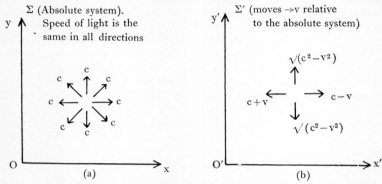

Figure 3.1. It is assumed that there is an absolute system Σ in which the speed of light in empty space is the same in all directions, as shown in (a). If it is assumed that the Galilean velocity transformations are correct, the speed of light in empty space should be different in different directions in the moving system Σ', as shown in (b).

absolute reference frame, it would be extremely unlikely that the earth was always at rest relative to it. Even if the earth happened to be at rest relative to the absolute system at one instant of time, due to the rotation of the earth about its axis and the orbital motion of the earth around the sun, the earth would not remain at rest relative to the absolute system at all periods of the day and year. Hence, if the Galilean transformations were correct, the speed of light in empty space should not be the same in all directions relative to the earth at all times. Hence, by measuring the speed of light in different directions of empty space relative to the earth, it should have been possible to estimate v, the speed of the earth relative to the absolute system in which the speed of light in empty space was the same in all directions. For example, the maximum and minimum speeds relative to the earth should have been $(c+v)$ and $(c-v)$ respectively, which would give c and v. An experiment of this type was attempted by Arago in 1810.

44

The most famous of all attempts was by Michelson and Morley in 1887. (An account of this experiment is given by Rosser[1 a].) All such optical experiments failed to find any differences in the speed of light in different directions and failed to establish the existence of any absolute system for the laws of optics. Michelson and Morley looked upon their experiment as a failure. According to the Galilean transformations they should have been able to determine the speed of the earth relative to the absolute system, but experimentally they found nothing, even though their apparatus was sensitive enough to detect the effect, if it were present. The way Einstein (reference 2) resolved this dilemma was to *assume* that the laws of optics obeyed the principle of relativity, even though this assumption meant abandoning the Galilean transformations and the concept of absolute time. Thus, according to Einstein, if one performed optical experiments inside an ocean liner (or spaceship) moving with uniform velocity relative to the earth, it would be impossible to determine its speed relative to the earth or any other inertial reference frame without looking at something external to the ocean liner (or spaceship). If the laws of optics satisfy the principle of relativity, it is impossible to detect any hypothetical absolute system by means of optical experiments and the whole idea of an absolute system for the laws of optics becomes superfluous. As his second postulate Einstein took the principle of the constancy of the speed of light, according to which the speed of light has the same numerical value in all inertial reference frames. For example, according to this postulate the speed of light should be equal to c in all directions relative to fig. 3.1 b as well as fig. 3.1 a. At first sight this tends to go against our common sense. However, common sense is almost always based on Newtonian mechanics. It will be illustrated in Chapter 4 how the principle of the constancy of the speed of light necessitates a revision of the classical ideas of absolute space and absolute time.

Some readers may like to repeat the argument illustrated in fig. 2.11 and discussed in § 2.8, replacing the electron by a pulse of light going at the limiting speed c relative to both Σ and Σ'.

3.3 *The laws of electricity and the principle of relativity*

The laws of electricity and magnetism, such as Coulomb's law for the force between two stationary point charges, the Biot-Savart law for the magnetic field of an electric current, Faraday's law of electromagnetic induction, etc., can be summarized in a concise way using Maxwell's equations. Physical optics can be interpreted in terms of electromagnetic waves using these same equations of electricity and magnetism. It can be shown that the speed of electromagnetic waves in empty space is $1/\sqrt{(\mu_0 \epsilon_0)}$, where μ_0 and ϵ_0 are the magnetic and the electric space constants respectively.

If it were assumed that the Galilean transformations were correct, then the generally accepted laws of electricity and magnetism would not agree with the principle of relativity. If the laws of electricity and magnetism did not obey the principle of relativity, then they could only hold in one absolute reference frame. The system in which the earth is at rest is not likely to be this absolute system. Thus, if the Galilean transformations were correct, it should have been possible to determine the speed of the earth relative to this absolute system by means of electrical experiments such as the Trouton–Noble experiment, an account of which is given by Rosser[1 b]. All such electrical experiments failed to determine the existence of any absolute system for the laws of electricity and magnetism. Einstein's way out of this dilemma also was to postulate that the laws of electricity and magnetism, which incorporate the laws of optics, obeyed the principle of relativity in inertial reference frames, even though this meant abandoning the Galilean transformations. If the laws of electricity and magnetism obey the principle of relativity, it is impossible to determine the existence of any absolute reference frame for the laws of electricity and magnetism. One could not determine the speed of an ocean liner (or spaceship) moving with uniform velocity relative to the earth by means of electrical or optical experiments confined to the inside of the ocean liner (or spaceship). In addition to the principle of relativity, as his second postulate Einstein could have postulated that the correct laws of classical optics and electromagnetism were Maxwell's equations. These postulates lead to the Lorentz transformations. However, Einstein chose the principle of the constancy of the speed of light as his second postulate, according to which the speed of light in empty space is the same in all inertial reference frames. This postulate simplified the development of the Lorentz transformations, and also enabled Einstein to discuss the measurement of the times of distant events. In his original paper, Einstein went on to show that the laws of electricity and magnetism (Maxwell's equations) do obey the principle of relativity, when the space and time co-ordinates are changed according to the Lorentz transformations.

3.4 *The principle of the constancy of the speed of light*

Some of the evidence in favour of the principle of the constancy of the speed of light will now be reviewed. If it is assumed that the accepted laws of electricity and magnetism (Maxwell's equations) obey the principle of relativity, then the principle of the constancy of the speed of light follows. A reader interested in a simple proof is referred to Rosser[1 c].

The principle of the constancy of the speed of light can be looked upon as a special case of the principle of the constancy of the limiting speed of particles, developed in § 2.8. It was illustrated in § 2.7(*b*) that light quanta (photons) can be treated as particles of zero rest mass

46

travelling at the speed of light. Photons can therefore be treated as particles travelling *at* the limiting speed. If the limiting speed of particles has the same numerical value in all inertial reference frames, then particles travelling at the limiting speed of $3 \cdot 0 \times 10^8$ metre per second must have the same speed of $3 \cdot 0 \times 10^8$ metre per second in all inertial reference frames, which for photons is just the principle of the constancy of the speed of light.

There is now direct experimental evidence in favour of the principle of the constancy of the speed of light. In very high energy nuclear disintegrations π^0-mesons are sometimes produced. These π^0-mesons have rest masses of 135 MeV$/c^2$, that is 264 times the rest mass of the electron. They are very unstable particles, and only live an average

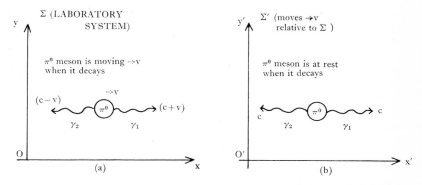

Figure 3.2. The π^0-meson is moving with uniform velocity v when it decays in the laboratory system Σ, as shown in (a). In Σ', the π^0-meson is at rest when it decays, so that the speeds of the photons γ_1 and γ_2 are both equal to c in Σ', as shown in (b). If the Galilean velocity transformations were correct, the speeds of γ_1 and γ_2 in the laboratory system Σ would be $(c+v)$ and $(c-v)$ respectively.

of about 10^{-16} second, when they are at rest, before decaying into two γ-rays, that is high energy light quanta (photons). In the inertial reference frame Σ' in which the decaying π^0-meson is at rest, the two γ-rays are emitted in opposite directions, so as to conserve linear momentum, as shown in fig. 3.2 b. In this reference frame the speeds of the γ-rays should be equal to the accepted value of $c = 2 \cdot 9979 \times 10^8$ metre per second for the speed of light emitted by a stationary light source.

Consider a π^0-meson moving with a velocity v relative to the laboratory system Σ, when it decays into two γ-rays, labelled γ_1 and γ_2, as shown in fig. 3.2 a. In the inertial frame Σ', moving with velocity v relative to Σ, the π^0-meson is at rest when it decays. Consider the special case shown in fig. 3.2 b, in which the two γ-rays (or light quanta) γ_1 and γ_2 are emitted parallel to and anti-parallel to the x' axis of Σ', with

47

speeds equal to c, the speed of light emitted by a stationary source. If the Galilean velocity transformations were correct, using equation (1.18), in the laboratory system Σ, we should have for the velocity of γ_1:

$$u_x = u_x' + v = c + v,$$

and for the velocity of γ_2:

$$u_x = u_x' + v = -c + v = -(c - v),$$

as shown in fig. 3.2 a. Thus, if the Galilean velocity transformations were correct, the speeds of the γ-rays (light quanta) arising from the decay of *moving* π^0-mesons should differ from c, the speed of light emitted by a stationary source.

In 1964 Alväger, Farley, Kjellman and Wallin measured the speeds of γ-rays (light quanta) from the decay of π^0-mesons of energy > 6 GeV relative to the laboratory. The π^0-mesons were produced by protons of momenta $19 \cdot 2$ GeV/c from the CERN proton synchrotron. In this case the speed of the light source, that is the π^0-mesons, relative to the laboratory was $0 \cdot 999\ 75c$, if the speed is calculated using the equations developed in Chapter 2. The measured speed of the γ-rays (light quanta) relative to the laboratory was $(2 \cdot 9977 \pm 0 \cdot 0004) \times 10^8$ metre per second. This was the same, within experimental error, as the accepted value of $2 \cdot 9979 \times 10^8$ metre per second for the speed of light emitted by a stationary light source. Fuller details of the experiment are given by Rosser[1 d]. These results show clearly that the speeds of light quanta (photons) emitted by a moving source are always equal to c, and the results illustrate vividly the inadequacy of the Galilean velocity transformations.

Some of the evidence in favour of the constancy of the speed of light has been reviewed in this section. The proof of the pudding is, however, in the eating. Some of the best evidence in favour of the principle of the constancy of the speed of light is the indirect evidence that the predictions made in Chapters 4 and 5, on the assumption that the principle of the constancy of the speed of light is correct, are invariably in agreement with the experimental results obtained in the laboratory. The principle of the constancy of the speed of light will be taken as *axiomatic* in Chapter 4.

References
1. Rosser, W. G. V., *Introductory Relativity* (Butterworths, London, 1967).
   (a) Appendix 1 (a); (b) Appendix 1 (b); (c) p. 299; (d) p. 167.
2. Einstein, A., *Annalen der Physik.*, **17** (1905), 891. This paper is translated into English in *Principles of Relativity* (Dover Publications, Inc., New York).

special relativity and the Lorentz transformations

4.1 *Postulates of special relativity*

EINSTEIN's two main postulates in the theory of special relativity were the principle of relativity and the principle of the constancy of the speed of light.

According to the *principle of relativity*, the laws of physics, including the laws of high speed particles and of optics, electromagnetism and nuclear physics, are the same in all *inertial reference frames*. The definition of an inertial reference frame is the same in the theory of special relativity as in Newtonian mechanics (cf. § 1.4). As an example of the scope of the principle of relativity in the theory of special relativity, consider a spaceship coasting along with *uniform velocity* with respect to the fixed stars. If an astronaut performed experiments in mechanics, optics, electromagnetism or nuclear physics inside the spaceship, according to the principle of relativity the laws the astronaut would ' derive ' on the basis of these experiments would be the same as if they were carried out in any other inertial reference frame such as the laboratory system. Without looking at anything external to the spaceship, the astronaut could not even say if the spaceship was moving, and he could not determine his speed relative to the earth. The astronaut should find no preferred directions in space.

According to the *principle of the constancy of the speed of light*, the speed of light in empty space has the same numerical value in all inertial reference frames. Thus, light should travel in straight lines and have the same speed in all directions of empty space in all inertial reference frames. The direct and indirect evidence in favour of the principle of the constancy of the speed of light was reviewed in § 3.4. This principle will be taken as *axiomatic* in this chapter. It will be shown that predictions made on the basis of this postulate are in agreement with the experimental results.

In this chapter radar methods will generally be used to determine the positions and times of events. Some people may prefer to use the principle of the constancy of the limiting speed of particles. Instead of radio signals, they can use fictitious space guns to shoot electrons back and forth with speeds *very*, *very* close to c. (It is, however, actually possible to use radio signals in practice.)

4.2 *Radar methods*

As an example of the use of radar methods, we shall consider the determination of the position and velocity of a spaceship moving directly away from the earth with uniform velocity, as shown in fig. 4.1. Let a directional antenna transmit a short radio pulse (or radar signal) at a time t_1, measured by a clock at rest in the radar station. This radar signal travels with a speed c in empty space. When it reaches the spaceship a small fraction of the radio waves are reflected in the backward direction, in the direction of the transmitter. These reflected signals can be received and amplified. Let the reflected signal return to the earth at a time t_2, measured by the clock at rest in the radar station. According to the principle of the constancy of the speed of light, the speed of the radar signal should be the same when the signal is going to the spaceship as when the reflected signal is

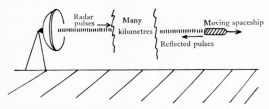

Figure 4.1. A directional antenna is used to transmit radar pulses which are reflected by the moving spaceship.

returning. Since the total time for the radar signal to go to the spaceship and back is $(t_2 - t_1)$, the time the radar signal takes to reach the spaceship is $\frac{1}{2}(t_2 - t_1)$ and the time it takes to return is $\frac{1}{2}(t_2 - t_1)$. Since it travels at a speed c, in a time interval of $\frac{1}{2}(t_2 - t_1)$, the radar signal travels a distance $\frac{1}{2}c(t_2 - t_1)$. Since it is transmitted at a time t_1 an observer in the radar station will *calculate* that the radar signal reached the spaceship at a time $t_1 + \frac{1}{2}(t_2 - t_1)$, that is at a time $\frac{1}{2}(t_2 + t_1)$. Hence, if the radar station is at the origin of a co-ordinate system, and the spaceship is moving along the x axis, the observer in the radar station will calculate that the radar signal was reflected by the spaceship at a time t at a distance x given by:

$$x = \frac{c}{2}(t_2 - t_1), \qquad (4.1)$$

$$t = \tfrac{1}{2}(t_2 + t_1). \qquad (4.2)$$

By repeating the experiment, the velocity of the spaceship can be determined. For example, let a radar signal be sent from the earth at 10.00 a.m. Let the reflected signal return to the earth 2 second later. According to equations (4.1) and (4.2) the co-ordinates and time of the event of reflection of the signal by the spaceship were:

$$x = \frac{c}{2}(t_2 - t_1) = \tfrac{1}{2} \times 3 \cdot 00 \times 10^8 \times 2 = 3 \cdot 00 \times 10^8 \text{ metre}$$

$$t = \tfrac{1}{2}(10.00.00 + 10.00.02) = 10.00.01 \text{ a.m.}$$

50

Thus the spaceship was a distance of $3 \cdot 00 \times 10^8$ metre from the earth at a time of 1 second after 10.00 a.m. If a signal transmitted at a time 10.20 a.m. returns to the earth 4 second later after reflection by the spaceship, according to equations (4.1) and (4.2) the spaceship was $6 \cdot 00 \times 10^8$ metre from the earth at a time 2 second after 10.20 a.m. In the time of 20 minute 1 second (1201 second), the spaceship travels a distance of $(6 \cdot 00 \times 10^8 - 3 \cdot 00 \times 10^8)$ or $3 \cdot 00 \times 10^8$ metre. Hence, if the spaceship is moving with uniform velocity the speed of the spaceship is $c/1201$ or ~ 250 kilometre per second.

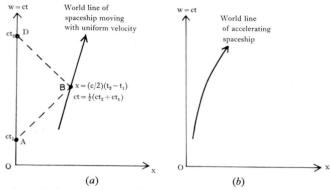

(a) (b)

Figure 4.2. (a) A space time diagram showing the determination of the position of a spaceship moving with uniform velocity. A radar pulse is transmitted from the radar base having co-ordinates $x = 0$, $w = ct_1$ at the world point A. It is reflected by the spaceship at the world point B at $x = \frac{1}{2}c(t_2 - t_1)$, $w = ct = \frac{1}{2}(ct_2 + ct_1)$, and received back at the radar base at the world point D at $x = 0$, $w = ct_2$. (b) The world line of an accelerating spaceship.

It is convenient to plot the positions of events such as the successive positions of a spaceship on a *displacement–time* curve. At present it will be assumed that all the events are on the x axis of the laboratory system Σ, and that the radar station is on the axis at $x = 0$. In the theory of special relativity, one is generally concerned with speeds comparable to the speed of light. For example, if a particle leaves the origin $x = 0$ at $t = 0$ and travels at a speed $c/2$, after a time $t = 1$ second, it is at $x = 1 \cdot 5 \times 10^8$ metre. With such high speed particles it is not convenient to plot distance in metres against time in seconds. It is a *convention* to multiply all times by c, the speed of light, before plotting the displacement–time curve. The symbol w will be used for the product ct. Therefore the co-ordinates of the particle at a time $t = 1$ second would be $x = 1 \cdot 5 \times 10^8$ metre, $w = ct = 3 \times 10^8$ metre. The quantity $w = ct$ has the dimensions of a length. It is also a convention to use $w = ct$ as ordinate and x as abscissa, as shown in fig. 4.2 a. Such a displacement–time diagram is called a *space–time* diagram. If the time of an event is t second, it will be shown as ct on space–time

51

diagrams, but it will generally be referred to as the time t in the text.

The displacement of a spaceship moving with uniform velocity v relative to the laboratory system Σ is shown in fig. 4.2 a. Such a displacement–time curve is known as a *world line*. The positions x, $w = ct$ of *events* in such a space–time diagram are known as *world points*. The reader should try to familiarize himself thoroughly with the definitions introduced in this section. Since light and radio signals go at a speed c, they travel a distance Δx equal to $c\Delta t$ in a time Δt. Hence the slopes of the world lines of light signals going along the $+x$ axis are $c\Delta t/\Delta x = 1$, so that the world lines of light signals going in the positive x direction are at $45°$ to the x and $w = ct$ axes. In fig. 4.2 a is shown the world line of the radar signal transmitted from the radar station at the point $x = 0$ of Σ at a time t_1 at world point A, which is reflected by the spaceship at world point B and returns to the radar station at $x = 0$ at a time t_2 at the world point D. The world line of the radar signal is at $45°$ to the x axis for the first half of the journey and at $135°$ to the x axis for the journey back from the spaceship to the radar station, as shown in fig. 4.2 a. If the spaceship is moving with uniform velocity relative to Σ, its world line is straight. If its velocity is v, in a time Δt it goes a distance $\Delta x = v\Delta t$ so that its inclination to the $w = ct$ axis is:

$$\tan \phi = \frac{\Delta x}{c\Delta t} = \frac{v\Delta t}{c\Delta t} = \frac{v}{c}. \tag{4.3}$$

Since for all particles of finite rest mass v, is less than c, the maximum value of $\tan \phi$ is 1, so that the inclinations of the world lines of all particles to the $w = ct$ axis must be less than $45°$. The world lines of light (radio) signals are at $45°$ to the $w = ct$ axis.

If the spaceship is accelerating relative to Σ, its world line is curved, as shown in fig. 4.2 b. Readers should draw a few space–time diagrams with world lines of particles and light signals so as to familiarize themselves with the method.

In this section we have described how one observer with a radar set can determine the positions and times of distant events. We shall now proceed to see how the radar measurements of the co-ordinates and times of distant events carried out by two observers moving with uniform velocity relative to each other can be related.

4.3 *Radar measurements carried out by two astronauts in relative motion*

(*a*) *Introduction*

Since many people have a mental bias in favour of the laboratory system, we shall consider two spaceships in outer space moving with different uniform velocities relative to the fixed stars. (If at any stage we want to use the laboratory system we can assume that one of the spaceships is at rest on the earth.) It will be assumed that astronaut John is on one spaceship and astronaut Mary is on the other. Both

John and Mary are inertial observers. Let John's spaceship be at rest at the origin of an inertial reference frame Σ, and let Mary's spaceship be at rest at the origin of an inertial reference frame Σ', as shown in fig. 4.3b. Let Mary's spaceship move with uniform velocity v relative to John's spaceship along the x axis of Σ, which is coincident with the x' axis of Σ', as shown in fig. 4.3 b, so that the inertial reference frame Σ' is moving with uniform velocity v relative to the inertial reference frame Σ. (The symbol Σ is the Greek capital letter sigma.) Let John and Mary construct identical clocks, and let them calibrate these clocks, using the frequency of the same atomic or nuclear transition. Let them also choose the same unit of length (such as the wavelength of the orange-red line of krypton-86), so that, according to the principle

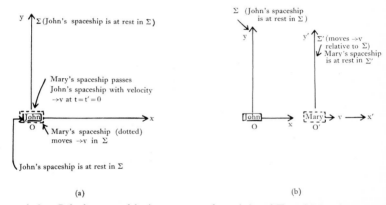

(a)   (b)

Figure 4.3. John's spaceship is at rest at the origin of Σ and Mary's spaceship is at rest at the origin of Σ'. Mary's spaceship moves with uniform velocity v relative to John's spaceship along the x axis of Σ, passing John's spaceship at $t = t' = 0$, as shown in (a). The x axis of Σ and the x' axis of Σ' are coincident.

of the constancy of the speed of light, they should measure the same numerical value c for the speed of light in all directions of empty space. Let John's and Mary's spaceships pass each other, when the clocks on John's and Mary's spaceships read $t = 0$ and $t' = 0$ respectively, as shown in fig. 4.3 a. (The origins of Σ and Σ' therefore coincide at $t = t' = 0$.)

(b) Radar determination of the co-ordinates of events

Let John send out a radar signal from his spaceship at a time t_1 on his clock. Let the radar signal be reflected at a distant *event* on the x axis of Σ, and let the reflected radar signal return to John's spaceship at a time t_2 on John's clock. If John assumes that the speed of radar signals (or light) is the same and equal to c in all direc-

53

tions, using equations (4.1) and (4.2), John will determine the position x and the time t of the event to be:

$$x = \frac{c}{2}(t_2 - t_1), \tag{4.4}$$

$$t = \tfrac{1}{2}(t_2 + t_1). \tag{4.5}$$

Equations (4.4) and (4.5) will be used with different symbols for x, t, t_1 and t_2 throughout this chapter to determine the positions of events relative to the inertial reference frame Σ in which John's spaceship is at rest.

Similarly, let Mary send out a radar signal from her spaceship at a time t_1' on her clock. Let the signal be reflected from an *event* on the x' axis of the inertial reference frame Σ', in which Mary's spaceship is at rest, and let the reflected signal reach Mary's spaceship at a time t_2' on Mary's clock. According to the principle of the constancy of the speed of light, which is being taken as axiomatic in this chapter, Mary must use the same value c for the speed of propagation of radar signals in all directions relative to the reference frame Σ', in which her spaceship is at rest. Hence, using equations (4.1) and (4.2) Mary will determine the position x' and the time t' of the event on the x' axis of Σ' to be:

$$x' = \tfrac{c}{2}(t_2' - t_1'), \tag{4.6}$$

$$t' = \tfrac{1}{2}(t_2' + t_1'). \tag{4.7}$$

These equations will be used throughout this chapter to determine the positions and times of events relative to the inertial reference frame Σ', in which Mary's spaceship is at rest at the origin.

Some readers may prefer to use science fiction space guns to send signals rather than use radio signals. These space guns would shoot electrons at such high speeds from John's and Mary's spaceships to events, and from events back to the spaceships, that the speeds of the electrons relative to both John and Mary would always be very, very, very close to c. Equations (4.4), (4.5), (4.6) and (4.7) and the analysis using radar methods in this chapter would then still be applicable. We shall continue to use radar methods in the discussion, since these methods can be used in practice to measure the times and positions of distant events.

(c) Relative speed of John's and Mary's spaceships

Let John send out a series of radar signals from his spaceship, which are reflected back to John's spaceship by Mary's spaceship. Let John use these signals to determine a series of values for the position of Mary's spaceship relative to his own spaceship at various times. John can then calculate the speed at which Mary's spaceship is moving away from his spaceship. Similarly, by transmitting radar signals

which are reflected by John's spaceship, Mary can determine the speed
at which John's spaceship is moving away from her spaceship. John
and Mary must agree on their speed of separation. If they did not
determine the same numerical value for this speed, it could only be due
to motion in one direction of empty space compared with motion in
the opposite direction of empty space. This is contrary to our assumption that there are no preferred directions in empty space, that is, that
space is isotropic. Hence John and Mary must agree on their speed
of separation, which will be denoted by v.

(d) The K-calculus

A space–time diagram (or displacement–time graph) will be used
to represent the positions and times of events relative to the inertial
reference frame Σ, in which *John's* spaceship is at rest. As described

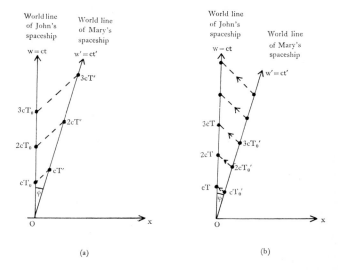

(a)   (b)

Figure 4.4. Space–time diagrams relative to the inertial reference frame Σ
in which John's spaceship is at rest. (a) John transmits radar signals at
times 0, T_0, $2T_0$, $3T_0$, etc. on his clock. These signals reach Mary's
spaceship at times 0, T', $2T'$, $3T'$, etc. on her clock, where $T'=KT_0$.
(b) Mary transmits radar signals at times 0, T_0', $2T_0'$, $3T_0'$, etc. on her
clock. These signals reach John's spaceship at times 0, T, $2T$, $3T$,
etc. on John's clock, where $T=KT_0'$.

in § 4.2, it is conventional to plot $w=ct$ against x, as shown in fig.
4.4 a, where x is the position and t is the time of the event, measured
by John using his radar set. The displacement of Mary's spaceship
relative to John's spaceship is shown in fig. 4.4 a. If Mary's spaceship
passes John's spaceship at $t=t'=0$, then Mary's world line Ow' is a
straight line through the origin. Since Mary's spaceship goes a

55

distance $x = vt = (v/c)ct = vw/c$ in a time t relative to John, Ow', that is Mary's world line, is at an angle ϕ to the Ow axis, where $\tan \phi$ is equal to v/c.

Throughout this chapter until § 4.10 *it will be assumed that all the world points and all the world lines are plotted relative to the inertial reference frame* Σ *in which John's spaceship is at rest using* Ox *and* Ow *as rectangular axes. We shall however take the liberty of labelling the times of events on Mary's world line* Ow' *as the times measured by Mary's clock.*

Let John send out radio signals at times $t = 0$, T_0, $2T_0$, $3T_0$, etc., as measured by John's clock. The time interval between the emission of successive signals is T_0 on John's clock. The world lines of these radio signals are at $45°$ to the x and $w = ct$ axes in fig. 4.4 a. Mary's spaceship coincides with John's spaceship, when John transmits his first signal at $t = 0$, so that Mary receives John's first signal at $t' = 0$ on her clock. Let Mary receive the radio signal transmitted by John at a time $t = T_0$ on John's clock at a time T' measured by Mary's clock. Now T' is greater than T_0, since Mary is moving away from John during the time interval the radio signal takes to travel from John's spaceship to Mary's spaceship. Let

$$T' = KT_0, \tag{4.8}$$

where $K > 1$. Since Mary is moving away from John, each successive signal transmitted by John has farther to go before reaching Mary than the previous one. However, if Mary is moving away from John with *uniform* velocity, each successive signal has the same extra distance to travel compared with the previous one. Hence the equally spaced radio signals transmitted by John at times 0, T_0, $2T_0$, etc., on his clock should reach Mary at times 0, T', $2T'$, etc. on her clock, as illustrated in fig. 4.4 a. For example, if John transmits radio signals at times 0, 1, 2, 3 second on his clock and the first two signals reach Mary's spaceship at times 0 and $1\frac{1}{2}$ second on Mary's clock, corresponding to $K = \frac{3}{2}$, and if Mary is moving away from John with *uniform* velocity, the successive signals from John should reach her at times 0, $1\frac{1}{2}$, 3, $4\frac{1}{2}$, 6 second, etc. on Mary's clock. This is really an example of the Doppler effect. A full discussion of the Doppler effect is given § 4.7.

Let Mary transmit radio signals at times $t' = 0$, 1, 2, 3 second, etc. measured by her clock, when she is moving away from John with the same uniform velocity as in the previous numerical example. According to the postulates of special relativity, all directions in space are equivalent and the speed of light is the same in all directions of empty space in all inertial reference frames. There is nothing to differentiate the inertial reference frame Σ' in which Mary is at rest from the inertial reference frame Σ in which John is at rest. John and Mary are just moving apart with uniform velocity relative to each other in outer space. Hence John should receive the radio signals Mary transmits at $t'=0$, 1, 2, 3 second on her clock at times $t=0$, $1\frac{1}{2}$, 3, $4\frac{1}{2}$ second on his own clock. Thus, if Mary measures the interval between John's equally spaced signals to be increased by a factor K when she is going away from John with uniform velocity v, John should measure the time interval between Mary's equally spaced signals to be increased by the same factor K, when John is moving away from Mary with the same uniform velocity v. If Mary transmits radio signals at times 0, T_0', $2T_0'$, $3T_0'$, etc. measured on her clock, these signals should reach John's spaceship at times 0, T, $2T$, $3T$, etc. measured by John's clock, where

$$T = KT_0', \qquad (4.9)$$

as illustrated in fig. 4.4 b. Equations (4.8) and (4.9) apply to the time intervals between successive signals. Provided Mary and John

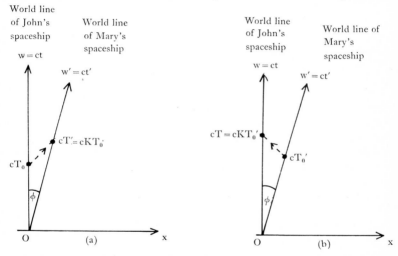

Figure 4.5. Mary's spaceship passes John's spaceship at $t=t'=0$ so that the world line of Mary's spaceship goes through the origin. (a) If John transmits a radio signal at a time T_0 on his clock it reaches Mary's spaceship at a time $T'=KT_0$ on Mary's clock. (b) If Mary transmits a radio signal at a time T_0' on her clock, it reaches John's spaceship at a time $T=KT_0'$ on John's clock. The space–time diagrams are relative to the inertial reference frame Σ in which John's spaceship is at rest.

pass each other at $t=t'=0$, then equations (4.8) and (4.9) can be used to relate the times of single radio signals sent from John to Mary at a time T_0 on John's clock, and from Mary to John at a time T_0' on Mary's clock.

Summarizing, if John transmits a radio signal to Mary at a time T_0 on his clock, and if this signal reaches Mary's spaceship at a time T' on Mary's clock, then

$$T' = KT_0. \qquad (4.10)$$

57

This is illustrated in fig. 4.5 *a*. Conversely, if we know the radio signal reaches Mary's spaceship from John's at a time T' on Mary's clock, then John must have transmitted the signal at a time T_0 on his clock, where

$$T_0 = T'/K. \tag{4.11}$$

If Mary transmits a radio signal to John's spaceship at a time T_0' on her clock, then the signal should reach John's spaceship at a time T on John's clock given by:

$$T = KT_0'. \tag{4.12}$$

This is illustrated in fig. 4.5 *b*. Conversely, if we know that Mary's signal reaches John at a time T on John's clock, Mary must have transmitted the signal at a time T_0' on her clock, where

$$T_0' = T/K. \tag{4.13}$$

Equations (4.10), (4.11), (4.12) and (4.13) are the basic formulae of the K-calculus. They hold for all values of T_0 and T_0', provided Mary passes John at $t = t' = 0$, and John and Mary continue to move apart with uniform velocity relative to each other. The reader should familiarize himself with these equations, and how they can be applied to the conditions illustrated in figs. 4.5 *a* and 4.5 *b*, *relating the times of transmission and reception of radio or light signals from John to Mary and from Mary to John respectively*. It is assumed that John and Mary coincide at $t = t' = 0$, so that Mary's world line goes through the origin. The value of the constant K will now be determined.

Let John transmit a radar signal at a time T_0 on his clock, as shown in fig. 4.6. Let this signal reach Mary at a time T' on Mary's clock where, according to equation (4.10):

$$T' = KT_0. \tag{4.14}$$

Let this radar signal be reflected by Mary's spaceship and return to John's spaceship at a time T on John's clock, as shown in fig. 4.6. From equation (4.12):

$$T = KT'.$$

Using equation (4.14):

$$T = K^2 T_0. \tag{4.15}$$

Using equations (4.4) and (4.5), John estimates that the radar signal was reflected from Mary's spaceship at:

$$x = \frac{c}{2}(T - T_0); \quad t = \tfrac{1}{2}(T + T_0).$$

If Mary's spaceship is moving at uniform speed v relative to John's

58

spaceship, since the two spaceships coincide at $t = 0$, John estimates that Mary's spaceship goes a distance x in a time t. Hence the speed of Mary's spaceship relative to John's spaceship is:

$$v = \frac{x}{t} = \frac{c(T - T_0)}{(T + T_0)}.$$

Using equation (4.15):

$$\frac{v}{c} = \frac{K^2 T_0 - T_0}{K^2 T_0 + T_0} = \frac{K^2 - 1}{K^2 + 1},$$

$$(v/c)K^2 + v/c = K^2 - 1,$$

$$K^2(1 - v/c) = 1 + v/c,$$

$$K = \sqrt{\left(\frac{1 + v/c}{1 - v/c}\right)}. \qquad (4.16)$$

This determines the factor K.

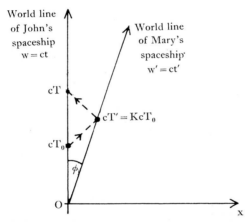

Figure 4.6. Determination of the constant K.   John transmits a radar signal at a time T_0 on his clock, which reaches Mary at a time $T' = KT_0$ on Mary's clock. The reflected signal reaches John's spaceship at time $T = KT' = K^2 T_0$ on John's clock.   John estimates the position of Mary's spaceship as $x = (c/2)[T - T_0]$ at $t = \frac{1}{2}[T + T_0]$.   Hence
$$v/c = x/ct = [T - T_0]/[T + T_0] = (K^2 - 1)/(K^2 + 1),$$ which gives
$K = [(1 + v/c)/(1 - v/c)]^{1/2}$.

Two important expressions involving K will now be developed:

$$K + \frac{1}{K} = \sqrt{\left(\frac{1 + v/c}{1 - v/c}\right)} + \sqrt{\left(\frac{1 - v/c}{1 + v/c}\right)}$$

$$= \frac{(1 + v/c) + (1 - v/c)}{\sqrt{[(1 - v/c)(1 + v/c)]}} = \frac{2}{\sqrt{(1 - v^2/c^2)}}.$$

59

Hence:

$$K + \frac{1}{K} = \frac{K^2 + 1}{K} = \frac{2}{\sqrt{(1 - v^2/c^2)}} = 2\gamma, \qquad (4.17)$$

where

$$\gamma = (1 - v^2/c^2)^{-1/2}. \qquad (4.18)$$

Similarly:

$$K - \frac{1}{K} = \frac{K^2 - 1}{K} = \frac{2v/c}{\sqrt{(1 - v^2/c^2)}} = \frac{2\gamma v}{c}. \qquad (4.19)$$

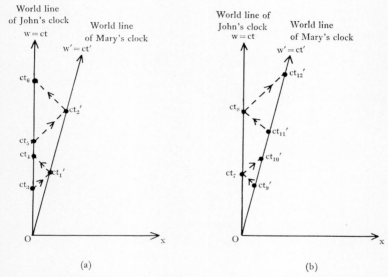

Figure 4.7. The comparison of the rates of two moving clocks using radar methods. (a) John uses radar methods to determine the times of the ticks at times t_1' and t_2' on Mary's clock. (b) Mary uses radar methods to determine the times of the ticks at times t_7 and t_8 on John's clock.

(e) *Time dilatation*

Let the times of two successive ticks on Mary's clock be at times t_1' and t_2' respectively, as measured by Mary's clock on her spaceship, which is moving with uniform velocity v relative to John's spaceship. Let John send a radar signal from his spaceship at a time t_3 on his clock, such that it is reflected from Mary's spaceship at the time of the first tick on Mary's clock, and let the reflected signal reach John's spaceship at a time t_4 on John's clock, as illustrated graphically in fig. 4.7 a. Let John transmit a second radar pulse at a time t_5 on his clock, such that it is reflected from Mary's spaceship at the time of the second tick on Mary's clock, and let the reflected signal return to John's spaceship at a time t_6 on John's clock as illustrated graphically in fig. 4.7 a. Mary estimates the time interval between the

60

two ticks on her clock to be $(t_2' - t_1')$. Using equations similar to equation (4.5), John estimates the times of the two ticks on Mary's clock to be:

$$t_1 = \tfrac{1}{2}(t_4 + t_3),$$

$$t_2 = \tfrac{1}{2}(t_6 + t_5).$$

Hence, using radar methods, John estimates the time difference between the two ticks on Mary's clock to be:

$$(t_2 - t_1) = \tfrac{1}{2}(t_6 + t_5) - \tfrac{1}{2}(t_4 + t_3). \qquad (4.20)$$

Using equation (4.11), we have:

$$t_3 = \frac{t_1'}{K}; \quad t_5 = \frac{t_2'}{K}.$$

Using equation (4.12), we have:

$$t_4 = Kt_1'; \quad t_6 = Kt_2'.$$

Substituting in equation (4.20):

$$(t_2 - t_1) = \tfrac{1}{2}(Kt_2' + t_2'/K) - \tfrac{1}{2}(Kt_1' + t_1'/K)$$

$$= \tfrac{1}{2}(K + 1/K)(t_2' - t_1').$$

But from equation (4.17) $(K + 1/K)$ equals 2γ. Hence:

$$(t_2 - t_1) = \gamma(t_2' - t_1') = \frac{(t_2' - t_1')}{\sqrt{(1 - v^2/c^2)}}. \qquad (4.21)$$

Equation (4.21) is the expression for time dilatation. Since

$$\gamma = (1 - v^2/c^2)^{-1/2}$$

is always greater than unity, using radar methods John measures the time interval between the two ticks on Mary's clock to be longer than Mary does. If $(t_2' - t_1')$ is one second on Mary's clock, by John's reckoning, using radar methods, the time between the two ticks on Mary's clock is γ second, whereas the time between two successive ticks on John's own clock would be 1 second. Hence, *using radar methods*, John measures Mary's clock, which is moving with uniform velocity v relative to him, to go at a slower rate than his own clock. The time interval between the two events (the two ticks on Mary's clock), as measured by Mary's clock, which coincides with both events, is called the *proper time interval* between the two events. John must use radar methods to measure the time interval between the two events. Even if John's clock coincided with the first tick on Mary's clock, due to their motion relative to each other, John's clock cannot coincide with the second tick on Mary's clock. The measurement of the rate of a moving clock, or the time interval between two events on a moving

61

spaceship necessitates the use of signals to transmit information to the 'stationary' observer.

Let the times of two successive ticks on John's clock be at t_7 and t_8, measured by John's clock, when John's and Mary's spaceships are moving apart with uniform speed v relative to each other. John estimates the time interval between the two events to be $(t_8 - t_7)$. Let Mary transmit radar signals at times t_9' and t_{11}' on her clock, such that they are reflected from John's spaceship at the times t_7 and t_8 respectively on John's clock, and let the reflected signals return to Mary's spaceship at times t_{10}' and t_{12}' respectively on Mary's clock, as illustrated graphically in fig. 4.7 b. Using equation (4.7), Mary estimates the time of the two events to be:

$$t_7' = \tfrac{1}{2}(t_{10}' + t_9'); \quad t_8' = \tfrac{1}{2}(t_{12}' + t_{11}').$$

Hence Mary measures the time intervals between the two ticks on John's clock to be $(t_8' - t_7')$, where

$$(t_8' - t_7') = \tfrac{1}{2}(t_{12}' + t_{11}') - \tfrac{1}{2}(t_{10}' + t_9').$$

Using equations (4.13) and (4.10):

$$t_9' = t_7/K; \quad t_{10}' = Kt_7; \quad t_{11}' = t_8/K; \quad t_{12}' = Kt_8.$$

Hence:

$$(t_8' - t_7') = \tfrac{1}{2}(Kt_8 + t_8/K) - \tfrac{1}{2}(Kt_7 + t_7/K)$$
$$= \tfrac{1}{2}(K + 1/K)(t_8 - t_7). \tag{4.22}$$

Using equation (4.17), equation (4.22) becomes:

$$(t_8' - t_7') = \gamma(t_8 - t_7) = \frac{(t_8 - t_7)}{\sqrt{(1 - v^2/c^2)}}. \tag{4.23}$$

In this case $(t_8 - t_7)$ is the proper time interval between the two events on John's spaceship, namely the two ticks on John's clock. Using radar methods, Mary measures the time interval between the two ticks on John's clock to be γ times what John does. Thus, using radar methods, Mary measures John's clock to go at a slower rate than her own clock. Time dilatation is perfectly reciprocal, provided Mary and John continue to move apart with uniform velocity relative to each other. Using radar methods, John measures Mary's clock, which is moving relative to him, to go at a slower rate than his own clock, whilst, using radar methods, Mary measures John's clock, which is moving relative to her, to go at a slower rate than her own clock. (The case of accelerating clocks will not be considered until Chapter 7.) The use of radar signals illustrates how time dilatation arises because the information is transmitted by signals travelling at finite speeds, and interpreted in a way consistent with the principle of the constancy of the speed of light.

(f) Simultaneity and length measurement*

In this section the K-calculus will be used to develop the expression for length contraction, and to show that spatially separated events, measured to be simultaneous in one inertial reference frame, are not measured to be simultaneous in an inertial reference frame moving relative to it. The algebra is a little more complicated than our previous

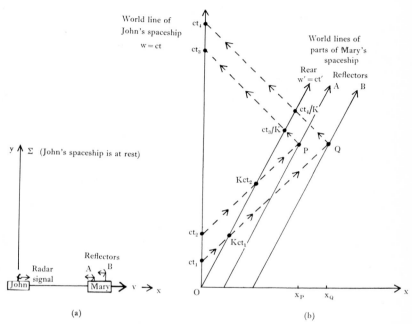

(a)

(b)

Figure 4.8. (a) John measures the separation of the reflectors A and B on Mary's spaceship, which is moving relative to John. (b) Space–time diagram showing the radar signals transmitted by John at times t_1 and t_2 on his clock, which after reflection at reflectors B and A respectively at world points Q and P return to John's spaceship at times t_4 and t_3 respectively on John's clock.

examples of the K-calculus, and some readers may prefer to move on to § 4.4, where the Lorentz transformations are developed. Using algebraic methods the Lorentz transformations can then be used to show the non-absolute nature of the simultaneity of spatially separated events, and to develop the expression for length contraction (cf. § 4.5).

It will be assumed that there are two reflectors, labelled A and B, on Mary's spaceship, as illustrated in fig. 4.8 a. As an example of the measurement of the length of a moving object, it will be assumed

63

that John uses radar methods to measure the separation of the reflectors A and B on Mary's spaceship, when Mary's spaceship is moving away from John's spaceship with uniform velocity v, as shown in fig. 4.8 a. The displacement of Mary's spaceship relative to the inertial reference frame Σ in which John's spaceship is at rest is shown in fig. 4.8 b. It will be assumed that the rear end of Mary's spaceship passes John's radar station at the origin of Σ at the time $t = t' = 0$, so that the world line of the rear of Mary's spaceship passes through the origin in fig. 4.8 b. The world lines of the reflectors A and B are parallel to the world line of the rear end of Mary's spaceship.

Let John send out a radar signal at a time t_1 on his clock, let this signal be reflected from reflector B on Mary's spaceship, and let the reflected signal return to John's spaceship at a time t_4 on John's clock, as shown graphically in fig. 4.8 b. The event of the reflection of this radar signal by reflector B on Mary's spaceship is represented by world point Q in fig. 4.8 b. Using equations (4.4) and (4.5), John estimates the co-ordinates of this event, relative to the inertial reference frame Σ in which he is at rest, to be

$$x_Q = \frac{c}{2} (t_4 - t_1), \qquad (4.24)$$

$$t_Q = \tfrac{1}{2} (t_4 + t_1). \qquad (4.25)$$

Let John transmit a second radar signal at a time t_2 on his clock, which is reflected by reflector A on Mary's spaceship, and let the reflected signal return to John's spaceship at a time t_3 on John's clock, as shown graphically in fig. 4.8 b. Using equations (4.4) and (4.5), John estimates the position and time of the reflection of the second radar signal by reflector A on Mary's spaceship (world point P in fig. 4.8 b) to be:

$$x_P = \frac{c}{2} (t_3 - t_2), \qquad (4.26)$$

$$t_P = \tfrac{1}{2} (t_3 + t_2). \qquad (4.27)$$

Now, if John wants to measure the separation of the reflectors A and B on Mary's spaceship, he is not going to measure the position of reflector A and then wait a time Δt before measuring the position of reflector B on Mary's spaceship since, in a time Δt, reflector B would have moved a distance $v\Delta t$ relative to John's spaceship. Obviously John must measure the positions of the reflectors A and B on Mary's spaceship at the same time on his reckoning, as measured by his radar methods. Hence John must arrange to transmit his radar signals such that

$$t_P = t_Q. \qquad (4.28)$$

64

If t_P equals t_Q, then the line joining world points P and Q in fig. 4.8 b must be parallel to the x axis. Substituting from equations (4.25) and (4.27) into equation (4.28):

$$\tfrac{1}{2}(t_4 + t_1) = \tfrac{1}{2}(t_3 + t_2).$$

Hence:

$$(t_4 - t_3) = (t_2 - t_1). \tag{4.29}$$

John estimates the separation of reflectors A and B on Mary's spaceship, which is moving with uniform velocity v relative to him, to be:

$$l = x_Q - x_P.$$

Using equations (4.24) and (4.26):

$$l = \frac{c}{2}(t_4 - t_1) - \frac{c}{2}(t_3 - t_2)$$

$$= \frac{c}{2}(t_4 - t_3) + \frac{c}{2}(t_2 - t_1).$$

Using equation (4.29):

$$l = x_Q - x_P = c(t_4 - t_3) = c(t_2 - t_1). \tag{4.30}$$

Let Mary be situated with her clock at the rear end of her spaceship. According to equation (4.10), the radar signals transmitted by John at times t_1 and t_2 on his clock should pass the rear end of Mary's spaceship at times Kt_1 and Kt_2 measured on Mary's clock, as marked on the world line Ow' in fig. 4.8 b. According to equation (4.13), the signals reflected at reflectors B and A at world points Q and P in fig. 4.8 b pass the rear of Mary's spaceship at times t_4/K and t_3/K respectively. Using equations (4.6) and (4.7), Mary will estimate the positions and times of the events at world points Q and P to be:

$$x_Q' = \frac{c}{2}\left(\frac{t_4}{K} - Kt_1\right); \quad t_Q' = \frac{1}{2}\left(\frac{t_4}{K} + Kt_1\right), \tag{4.31}$$

$$x_P' = \frac{c}{2}\left(\frac{t_3}{K} - Kt_2\right); \quad t_P' = \frac{1}{2}\left(\frac{t_3}{K} + Kt_2\right). \tag{4.32}$$

Notice t_Q' is not equal to t_P'. However, it does not matter when Mary measures the positions of the reflectors A and B on her spaceship as they are not moving relative to her. She can lay out a ruler at *rest* relative to her spaceship. She can look at the position of reflector A on the ruler and come back half an hour later to look at the position of reflector B. Since nothing moves relative to her, she would measure the correct length. Hence Mary estimates the distance l_0 between the

3   65

reflectors A and B on her spaceship to be:

$$l_0 = x_Q' - x_P' = \frac{c}{2}\left(\frac{t_4}{K} - Kt_1\right) - \frac{c}{2}\left(\frac{t_3}{K} - Kt_2\right)$$

$$= \frac{c}{2}\left\{\frac{(t_4 - t_3)}{K} + K(t_2 - t_1)\right\}.$$

From equation (4.29), $(t_4 - t_3)$ equals $(t_2 - t_1)$, so that

$$l_0 = \frac{c}{2}\left\{\frac{1}{K} + K\right\}(t_4 - t_3) = \frac{c}{2}\left\{\frac{1}{K} + K\right\}(t_2 - t_1).$$

Using equation (4.17):

$$l_0 = x_Q' - x_P' = \gamma c(t_4 - t_3) = \gamma c(t_2 - t_1). \qquad (4.33)$$

But from equation (4.30):

$$c(t_4 - t_3) = c(t_2 - t_1) = x_Q - x_P = l.$$

Hence equation (4.33) becomes:

$$l_0 = \gamma l$$

or

$$l = l_0 \sqrt{(1 - v^2/c^2)}. \qquad (4.34)$$

This is the expression for length contraction, or the Lorentz contraction as it is sometimes called. The length of an object measured in the inertial reference frame in which it is at rest is known as the *proper length* of the object. The length l_0 between the reflectors A and B measured by Mary is the proper length in this case. According to equation (4.34), John would measure the length of the moving object (the separation of reflectors A and B on Mary's spaceship) to be less than the proper length, that is, less than the length measured by Mary. Thus, using radar methods, John would *measure* all bodies moving relative to him to be Lorentz contracted in their direction of motion relative to him. Similarly, Mary would measure John's spaceship, which is moving relative to her, to be Lorentz contracted. If $v \ll c$, then $l \simeq l_0$, so that the concept of absolute length is a satisfactory *approximation* in the context of Newtonian mechanics.

From equations (4.31) and (4.32):

$$t_Q' - t_P' = \frac{1}{2}\left(\frac{t_4}{K} + Kt_1\right) - \frac{1}{2}\left(\frac{t_3}{K} + Kt_2\right)$$

$$= \tfrac{1}{2}(t_4 - t_3)/K - \tfrac{1}{2}K(t_2 - t_1).$$

Since from equation (4.29):

$$(t_4 - t_3) = (t_2 - t_1),$$

$$t_Q' - t_P' = \tfrac{1}{2}(t_4 - t_3)(1/K - K) = -\tfrac{1}{2}(t_4 - t_3)(K - 1/K).$$

66

Substituting for $(K - 1/K)$ from equation (4.19):

$$t_Q' - t_P' = -\tfrac{1}{2}(t_4 - t_3)\frac{2\gamma v}{c} = -\frac{\gamma v}{c^2}c(t_4 - t_3).$$

Substituting from equation (4.30) for $c(t_4 - t_3)$:

$$t_Q' - t_P' = -\frac{\gamma v}{c^2}(x_Q - x_P). \qquad (4.35)$$

It can be seen from fig. 4.8 b that, relative to the reference frame Σ in which John's spaceship is at rest, $x_Q > x_P$. Hence from equation (4.35) it follows that t_Q' is less than t_P', so that on Mary's reckoning the event at world point Q, when the radar signal was reflected at reflector B, occurred before the event at world point P when the radar signal was reflected by reflector A. Mary will therefore conclude that John measured the position of the reflector B on her spaceship before he measured the position of reflector A, though John measured these events (at world points P and Q in fig. 4.8 b) to be simultaneous. Thus the simultaneity of spatially separated events is not absolute, as it was assumed to be in Newtonian mechanics, though if $v \ll c$, $t_Q' \simeq t_P'$, and the concept of absolute time is a satisfactory *approximation* in the context of Newtonian mechanics. According to the theory of special relativity, if two events are observed by measurement to be simultaneous in one inertial reference frame, they are not observed by measurement to be simultaneous in an inertial reference frame moving relative to it. Mary will conclude that it was no wonder that John measured the separation of the reflectors A and B on her spaceship to be less than she did. On her reckoning, based on *her* measurements, Mary will say that John measured the position of reflector B on her spaceship at a time t_Q' on her reckoning, and then waited for a time interval of $(t_P' - t_Q')$, during which time John was moving in the direction from B to A *relative to Mary* before John measured the position of reflector A on Mary's spaceship. Thus the measurement of the lengths of moving objects depends intimately on the measurement of the times of distant events.

4.4 *The Lorentz transformations*

It will again be assumed that John's spaceship is at the origin of the inertial reference frame Σ, that Mary's spaceship is at the origin of the inertial reference frame Σ', that Mary's spaceship is moving with uniform velocity v relative to John's spaceship, such that Σ' is moving with *uniform* velocity v relative to Σ along their common x axis, as shown previously in fig. 4.3 b. Let Mary's spaceship pass John's spaceship at $t = t' = 0$, so that the origins of Σ and Σ' coincide at $t = t' = 0$. Let an *event* occur on the x axis of Σ, at a point farther away from John's spaceship than Mary's spaceship. Let John measure the position and time of this event, using radar methods. Let John use the symbols x and t to denote the position and time of this event. It follows from equations (4.4) and (4.5) that John must have transmitted the radar signal at a time $t_1 = (t - x/c)$ on his clock and received the signal reflected at the event back at a time $t_2 = (t + x/c)$ on his clock, if he calculates the position of the event to be x at a time t. (Check: From equation (4.5) the time of the event is

$$\tfrac{1}{2}(t_2 + t_1) = \tfrac{1}{2}\{(t + x/c) + (t - x/c)\} = t.$$

Similarly, using equation (4.4), the position of the event is $(c/2)(t_2 - t_1) = x$.) Let John's radar signal pass Mary's spaceship at a time $(t' - x'/c)$ on Mary's clock, and let the signal reflected by the

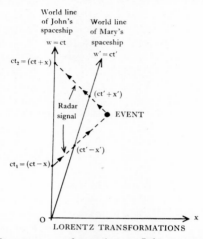

Figure 4.9. The Lorentz transformations. John transmits a radar signal at a time $(t - x/c)$ on his clock. The radar signal passes Mary's spaceship at a time $(t' - x'/c)$ on her clock, and, after reflection at the event, passes Mary's spaceship at a time $(t' + x'/c)$ on Mary's clock, before reaching John's spaceship at a time $(t + x/c)$ on John's clock.

event pass Mary at a time $(t' + x'/c)$ on Mary's clock, as illustrated in fig. 4.9, such that using equations (4.6) and (4.7), Mary estimates the position of the event to be at x' at a time t' relative to Σ'.

Treating the time from 0 to $(t - x/c)$ on John's clock as T_0, and the time from 0 to $(t' - x'/c)$ on Mary's clock as T', from equation (4.10) we have after multiplying both sides by c:

$$ct' - x' = K(ct - x) = Kct - Kx. \tag{4.36}$$

Treating the interval between 0 and $(t' + x'/c)$ on Mary's clock as T_0' and the interval between 0 and $(t + x/c)$ on John's clock as T, from equation (4.13):

$$ct' + x' = \frac{(ct + x)}{K} = \frac{ct}{K} + \frac{x}{K}. \tag{4.37}$$

68

Subtracting equation (4.36) from equation (4.37):

$$2x' = x\left(\frac{1}{K} + K\right) - ct\left(K - \frac{1}{K}\right).$$

From equation (4.17) $(K + 1/K)$ equals 2γ and from equation (4.19) $(K - 1/K)$ equals $2\gamma v/c$. Hence:

$$2x' = 2\gamma x - ct \, 2\gamma v/c$$

or

$$x' = \gamma(x - vt), \qquad (4.38)$$

where

$$\gamma = 1/\sqrt{(1 - v^2/c^2)}. \qquad (4.39)$$

Adding equations (4.36) and (4.37) gives:

$$2ct' = ct(1/K + K) - x(K - 1/K).$$

Using equations (4.17) and (4.19), we obtain:

$$2ct' = ct2\gamma - x2\gamma v/c.$$

Dividing both sides by $2c$ gives:

$$t' = \gamma(t - vx/c^2). \qquad (4.40)$$

Equations (4.38) and (4.40) are the Lorentz transformations relating the positions x and x' and times t and t' of the event, measured by John and Mary respectively using radar methods. The use of radar methods makes it clear that the Lorentz transformations relate the measured positions and times of events. It is left as an algebraic exercise for the reader to rearrange equations (4.38) and (4.40) to show that

$$x = \gamma(x' + vt'); \quad t = \gamma(t' + vx'/c^2). \qquad (4.41)$$

These are the inverse transformations. Notice the inverse transformations can be obtained from equations (4.38) and (4.40) by changing primed into unprimed quantities and unprimed quantities into primed quantities and replacing v by $-v$. This procedure works for all relativistic transformations.

The measurement of lengths in a direction perpendicular to the common x axis, that is, perpendicular to the direction of motion of Mary's spaceship relative to John's spaceship will now be discussed. Let Mary place her radar set on the x' axis of the inertial reference frame Σ', in which her spaceship is at rest, as shown in fig. 4.10 b. Let Mary use radar methods to measure the distance y' between her radar set on the x' axis and a reflector R on her spaceship, as shown in fig. 4.10 b. Let the line joining Mary's radar set and the reflector

69

R be perpendicular to the direction in which John's spaceship is moving with velocity v along the negative x' axis of Σ', as shown in fig. 4.10 b. Let Mary transmit a radar signal at a time t_1' on her clock, and let this signal be reflected by the reflector R, and let the reflected signal return to Mary's radar set at a time t_2' on Mary's clock, as shown in fig. 4.10 b. Since the radar signal travels a total distance $2y'$ to the reflector R and back in a time interval $(t_2' - t_1')$ relative to the inertial reference frame Σ' in which Mary's spaceship is at rest, Mary would conclude that in Σ':

$$2y' = c(t_2' - t_1').\qquad(4.42)$$

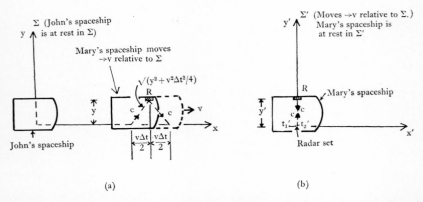

(a)   (b)

Figure 4.10. The measurement of a length perpendicular to the direction of relative motion. (a) John's interpretation of Mary's measurements. (b) Mary measures y' the distance between the x' axis and a reflector R on the y' axis of Σ', using radar methods.

Using the K-calculus, it was shown in § 4.3 (e) [compare equation (4.21)] that, using radar methods, John will measure the time interval Δt between the ticks at t_1' and t_2' on Mary's clock, when the radar signal left and returned to the x' axis respectively to be: $\Delta t = \gamma(t_2' - t_1')$. Hence, using equation (4.42):

$$\Delta t = 2\gamma y'/c.\qquad(4.43)$$

Relative to the inertial reference frame Σ in which John's spaceship is at rest, Mary's spaceship is moving with uniform velocity v along the $+x$ axis, and, relative to Σ, the radar signal takes the path shown in fig. 4.10 a. In the time interval Δt measured by John, Mary's spaceship goes a distance $v\Delta t$ relative to John's spaceship. By symmetry, Mary's spaceship travels a distance $v\Delta t/2$ in the time the radar signal takes to go from the transmitter on the x' axis to the reflector R. Let the reflector R be at a perpendicular distance y from the x axis, measured relative to Σ, the reference frame in which John's spaceship

70

is at rest. In fig. 4.10 a, it follows from Pythagoras' theorem that, relative to Σ, the distance from the point of emission of the radar signal to the reflector R is $\sqrt{[y^2+(v\Delta t/2)^2]}$. The total distance travelled by the radar signal relative to Σ is $2\sqrt{[y^2+(v\Delta t/2)^2]}$. Since light travels with a speed c during the time Δt it takes to cover this distance, relative to the reference frame Σ shown in fig. 4.10 a, in which John's spaceship is at rest, we have:

$$c\Delta t = 2\sqrt{(y^2+v^2\Delta t^2/4)}.$$

Squaring both sides:

$$c^2\Delta t^2 = 4(y^2+v^2\Delta t^2/4) = 4y^2+v^2\Delta t^2.$$

Substituting $2\gamma y'/c$ for Δt from equation (4.43), we have:

$$\frac{c^2 4\gamma^2 y'^2}{c^2} = 4y^2 + \frac{v^2 4\gamma^2 y'^2}{c^2}.$$

Rearranging:

$$\gamma^2 y'^2(1-v^2/c^2) = y^2.$$

Since γ^2 equals $1/(1-v^2/c^2)$, taking the square root we have:

$$y' = y.$$

Hence John and Mary should agree on the measurements of lengths perpendicular to the x and x' axes, that is perpendicular to the directions of their motions relative to each other. Hence in general, if Σ' moves with uniform velocity v relative to Σ along a common x axis:

$$y' = y, \tag{4.44}$$

$$z' = z. \tag{4.45}$$

Collecting the Lorentz transformations, we have:

$$x' = \gamma(x-vt), \qquad x = \gamma(x'+vt'),$$
$$y' = y, \qquad y = y',$$
$$z' = z, \qquad z = z',$$
$$t' = \gamma(t-vx/c^2), \qquad t = \gamma(t'+vx'/c^2).$$

The Lorentz transformations relate the co-ordinates and times of *events* in one inertial co-ordinate system Σ to the co-ordinates and time of the *same* event measured relative to another inertial co-ordinate system Σ', which is moving relative to Σ along their common x axis, and whose origin coincides with the origin of Σ at $t=t'=0$. The reader should try to become as familiar as possible with these equations. A few simple problems are given at the end of this chapter.

71

If we let $v/c = \sin\theta$,

$$\gamma = \frac{1}{\sqrt{(1 - v^2/c^2)}} = \frac{1}{\sqrt{(1 - \sin^2\theta)}} = \frac{1}{\sqrt{(\cos^2\theta)}} = \sec\theta.$$

Hence γ can be calculated, using trigonometrical tables. For example, if v/c is $0 \cdot 8000$, using four-figure tables one finds θ is $53^\circ 8'$ and sec θ is $1 \cdot 6668$. By direct calculation γ is $5/3$.

Notice if $c \to \infty$, the Lorentz transformations become:

$$x' = (x - vt); \quad y' = y; \quad z' = z; \quad t' = t$$

in agreement with the Galilean transformations developed in Chapter 1.

So far in this chapter, radar methods have been used when discussing the determination of the co-ordinates and times of distant events. This has the advantage of making it clear from the outset that one is dealing with the *measured* times of distant events and that one must allow for the propagation times of signals. It is not always convenient to use radar methods. For example, if an astronaut were in outer space in a spaceship at *rest* relative to the earth, it would be pointless to use radar methods from a base station on the earth to measure the times of events inside the spaceship. It is far more convenient for the astronaut to use his own clock. The question arises as to how the astronaut can synchronize his clock to keep the same time as a master clock on the earth. If the astronaut takes his clock with him, one cannot be sure that his clock is not affected by the transportation. Einstein's prescription for synchronizing distant clocks will be illustrated by an imaginary radio conversation between the astronaut and his base station. The astronaut could ask his base station on earth to send him a radio signal. On receipt of the radio signal from earth, the astronaut should send a radio signal back to the earth without time delay. The astronaut should also note the time on his clock when he received the radio signal from earth. If the radio signal left the earth at a time t_1 on the master clock at the base station on the earth and the signal from the astronaut was received back at a time t_2 on the master clock on the earth, since the speed of light in empty space is the same in all directions, the controller at the base station will estimate that the radio signal reached the astronaut at a time $\frac{1}{2}(t_1 + t_2)$. This information can be transmitted to the astronaut, who can then set his clock accordingly. This procedure is similar to the radar method. Clocks synchronized in this way, and distributed throughout space could be used to determine the times of distant events when and where they occur. In the laboratory it is convenient to send messages from events as electrical signals along electric cables. It does not matter what method is used to measure the times of distant events, provided it is done in a way consistent with the theory of special relativity. In future, when applying the Lorentz transformations, we shall not always specify precisely how the measurements are carried out, but will assume that they are carried out and interpreted in a way consistent with the theory of special relativity.

4.5 *Applications of the Lorentz transformations*

In §§ 4.3 (*e*) and (*f*) the expressions for time dilatation, length contraction and non-absolute simultaneity were developed directly using radar methods and the K-calculus. It will now be illustrated how these results can be developed using the Lorentz transformations.

(*a*) *Non-absolute simultaneity and length contraction*

Consider the example discussed in § 4.3 (*f*) and illustrated previously in fig. 4.8 *a*. John measures the positions of two reflectors A and B on Mary's spaceship, when Mary's spaceship is moving away from John, at the same time *t* on John's reckoning. If John measures the positions of the reflectors A and B to be x_P and x_Q, at the same time $t(=t_P=t_Q)$. Using the Lorentz transformations, relative to Σ', the reference frame in which Mary's spaceship is at rest, we have:

$$t_P' = \gamma\left(t - \frac{v}{c^2}x_P\right); \quad t_Q' = \gamma\left(t - \frac{v}{c^2}x_Q\right).$$

Hence:

$$t_Q' - t_P' = -\gamma\frac{v}{c^2}(x_Q - x_P). \tag{4.46}$$

This is in agreement with equation (4.35) and shows that events measured to be simultaneous in one inertial reference frame are not measured to be simultaneous in a reference frame moving with uniform velocity relative to it. From the Lorentz transformations, in Σ' we have:

$$x_P' = \gamma(x_P - vt); \quad x_Q' = \gamma(x_Q - vt).$$

Hence:

$$x_Q' - x_P' = \gamma(x_Q - x_P),$$
$$l_0 = \gamma l,$$
$$l = l_0\sqrt{(1 - v^2/c^2)}. \tag{4.47}$$

This is the same as equation (4.34). It is the expression for length contraction; $l_0 = x_Q' - x_P'$ is the proper length, that is the separation of the reflectors A and B on Mary's spaceship measured in the inertial frame Σ' in which Mary's spaceship is at rest, and $l = x_Q - x_P$, is the length measured by John, relative to whom the object (Mary's spaceship) is moving with uniform velocity *v*.

The phenomenon of length contraction is reciprocal. For example, let an object be at rest on the *x* axis of Σ, and let the positions of its

73

extremities relative to Σ be x_1 and x_2, so that its proper length is $(x_2 - x_1)$. Let the positions of its ends be measured to be at the points x_1' and x_2' at the *same* time t' in Σ'. From the Lorentz transformations:

$$x_1 = \gamma(x_1' + vt'); \quad t_1 = \gamma(t' + vx_1'/c^2),$$
$$x_2 = \gamma(x_2' + vt'); \quad t_2 = \gamma(t' + vx_2'/c^2).$$

Notice t_1 is not equal to t_2, showing that the events are not simultaneous in Σ. Subtracting:

$$(x_2 - x_1) = \gamma(x_2' - x_1')$$

or

$$(x_2' - x_1') = (x_2 - x_1)\sqrt{(1 - v^2/c^2)}.$$

In this case $(x_2 - x_1)$ is the proper length of the object, which is at rest relative to Σ, and $(x_2' - x_1')$ its length when it is moving with velocity $-v$ relative to Σ'.

(b) Time dilatation

The example discussed in § 4.3 (*e*) will now be reconsidered using the Lorentz transformations. Consider the two ticks at times t_1' and t_2' on Mary's clock, as illustrated previously in fig. 4.7 *a*. In Σ' the position of Mary's clock is $x' = 0$. Using the Lorentz transformations, the times of the two ticks on Mary's clock relative to Σ are, since $x' = 0$:

$$t_1 = \gamma(t_1' + vx'/c^2) = \gamma t_1',$$
$$t_2 = \gamma(t_2' + vx'/c^2) = \gamma t_2'.$$

Hence:

$$(t_2 - t_1) = \gamma(t_2' - t_1') = \frac{t_2' - t_1'}{\sqrt{(1 - v^2/c^2)}}. \tag{4.48}$$

Equation (4.48) is in agreement with equation (4.21). The time interval $(t_2' - t_1')$ is the proper time interval measured by *one* clock (Mary's) which coincides with both events. John must use two synchronized clocks (or radar methods) to measure $(t_2 - t_1)$ since, if a clock at rest relative to Σ coincides with Mary's clock at a time t_1' on her clock, it does not coincide with Mary's clock at the later time t_2' on Mary's clock, since Mary's clock is moving relative to Σ.

Since John's clock is at $x = 0$ in Σ, relative to the reference frame Σ' in which Mary's spaceship is at rest, the ticks at times t_7 and t_8 on John's clock (cf. fig. 4.7 (*b*)) are at times

$$t_7' = \gamma(t_7 - vx/c^2) = \gamma t_7$$
$$t_8' = \gamma(t_8 - vx/c^2) = \gamma t_8$$

74

relative to Σ'. Subtracting,

$$(t_8' - t_7') = \gamma(t_8 - t_7) \qquad (4.49)$$

This is in agreement with equation (4.23), showing that time dilatation is reciprocal. The time interval $(t_8 - t_7)$ is the proper time interval in this case, since it is measured by one clock (John's) which coincides with both events. Equations (4.48) and (4.49) show that proper time intervals in either Σ or Σ' are measured to be longer in the other reference frame. Some of the experimental evidence in favour of time dilatation will now be discussed.

4.6 *Experimental verification of time dilatation**

According to the law of radioactive decay, if we have N_0 radioactive atoms at *rest* at $t = 0$, the number N left after a time t is given by:

$$N = N_0 \exp{(-\lambda t)},$$

where λ is the decay constant. In a time equal to the half life $T_{1/2} = 0 \cdot 693/\lambda$, on average, half the radioactive atoms will have decayed. In a further time $T_{1/2}$, on average half of the remaining radioactive atoms will have decayed, etc. It can be shown that the average or mean time T_0 which a radioactive atom lives before it decays is equal to $1/\lambda$. Hence the law of radioactive decay can be rewritten:

$$N = N_0 \exp{(-t/T_0)},$$

where T_0 is the *average* or *mean life* of the radioactive atoms. The half life $T_{1/2}$ is equal to $0 \cdot 693 \, T_0$. In high energy nuclear physics it is conventional to use the mean life T_0, rather than λ and $T_{1/2}$.

Consider a box of radioactive atoms inside a spaceship which is at rest at the origin of the inertial frame Σ', and which is moving with uniform velocity v relative to the laboratory system Σ. Let the spaceship pass the origin of the laboratory system Σ at $t = 0$ and $t' = 0$ relative to Σ and Σ' respectively. Let there be a total of N_0 radioactive atoms in the spaceship at the time $t' = 0$. According to the law of radioactive decay, the number N left in the spaceship after a time t', relative to the spaceship (Σ'), is

$$N = N_0 \exp(-t'/T_0) \qquad (4.50)$$

where T_0 is the average time a radioactive particle lives relative to the inertial reference frame in which it is at rest [the spaceship (Σ') in this instance]. Relative to the laboratory system Σ, the radioactive atoms are moving with uniform velocity v along the x axis. Corresponding to $x' = 0$, $t' = t'$ relative to the spaceship (Σ'), according to the Lorentz transformations, in the laboratory system Σ we have:

$$x = \gamma(x' + vt') = \gamma vt'; \quad \text{or} \quad t' = x/\gamma v, \qquad (4.51)$$

$$t = \gamma(t' + vx'/c^2) = \gamma t'; \quad \text{or} \quad t' = t/\gamma. \qquad (4.52)$$

75

Substituting for t' in equation (4.50), using equations (4.52) and (4.51), we have:

$$N = N_0 \exp\left(-t/\gamma T_0\right) \qquad (4.53)$$

and

$$N = N_0 \exp\left(-x/\gamma v T_0\right). \qquad (4.54)$$

Hence, relative to the laboratory system Σ, the radioactive particles live for an average time γT_0, and travel an average distance $\gamma v T_0$ before decaying. Equations (4.53) and (4.54) have been confirmed by experiments with π-mesons and with μ-mesons. The π-meson is

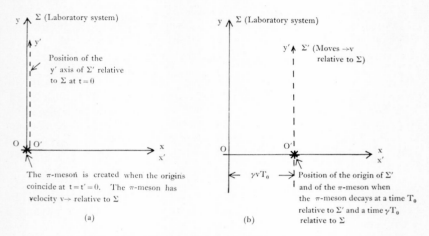

Figure 4.11. (a) A π-meson is created at the origin of Σ' at $t = t' = 0$ when the origins of Σ and Σ' coincide. If the π-meson moves with uniform velocity v relative to Σ, it remains at the origin of Σ'. (b) The position of the π-meson when it decays.

associated with the forces between nucleons (protons and neutrons) inside atomic nuclei, and π-mesons are produced in collisions between high energy nucleons [cf. §§ 2.7 (d) and 5.5 (b)]. The mass of the π-meson is $139 \cdot 6 \text{ MeV}/c^2$. The π-meson is unstable, decaying into a μ-meson and a neutrino, the mean lifetime of π-mesons being $2 \cdot 55 \times 10^{-8}$ second in the inertial reference frame in which they are at rest. The mass of a μ-meson is $105 \cdot 7 \text{ MeV}/c^2$, it decays into an electron plus two neutrinos and its mean life is $2 \cdot 2 \times 10^{-6}$ second in the inertial reference frame in which it is at rest.

The decays of moving radioactive atoms are distributed statistically in the way given by equations (4.53) and (4.54). For purposes of discussion we shall consider the special case of a π-meson, which is moving with velocity v relative to the laboratory system Σ, and which lives for a time T_0 in the inertial reference frame Σ' in which it is at

76

rest. Let the π-meson be created at $t=t'=0$ at the instant when the origins of Σ and Σ' coincide, as shown in fig. 4.11 a. The π-meson remains at rest at the origin of Σ' and decays at $x'=0$ at a time $t'=T_0$ relative to Σ' as shown in fig. 4.11 b. Relative to the laboratory system Σ the π-meson would be created at a time $t=0$ at $x=0$ and would decay at

$$x = \gamma(x' + vt') = \gamma v T_0,$$

$$t = \gamma(t' + vx'/c^2) = \gamma T_0.$$

The time T_0 in Σ' is the proper time interval between the event of the creation and the event of the decay of the π-meson at the origin of Σ'. Relative to the laboratory system Σ, the π-meson travels with a velocity v for a time γT_0, covering a distance $\gamma v T_0$ as illustrated in fig. 4.11 b. Hence, relative to the laboratory system Σ, we have time dilatation; that is, π-mesons should be measured to live longer when they are moving relative to the laboratory, compared with when they are at rest relative to the laboratory. Experiments have shown that the mean lifetime of π-mesons when they are at rest in the laboratory system Σ is $T_0 = 2 \cdot 55 \times 10^{-8}$ second. The maximum or limiting speed of particles is the speed of light $c = 3 \times 10^8$ metre per second. Hence, if there were no time dilatation, π-mesons would only go an average distance of $3 \times 10^8 \times 2 \cdot 55 \times 10^{-8} = 7 \cdot 6$ metre, before decaying, even if they travelled at the speed of light.   In 100 metre the intensity of π-mesons would decrease to $\exp(-100/7 \cdot 6)$ or $\sim 2 \times 10^{-6}$ of the original intensity, if there were no time dilatation. Allowing for time dilatation, π-mesons of total energy $4 \cdot 5$ GeV, $u = 0 \cdot 9995c$, $\gamma \simeq 32$, should live a mean time $\gamma T_0 = 81 \times 10^{-8}$ second and travel a mean distance $\gamma v T_0 = 244$ metre before decaying. Experiments at CERN have confirmed that π-mesons of energy $4 \cdot 5$ GeV travel a mean distance of ~ 250 metre before decaying. In fact, at CERN the π-meson detectors, such as bubble chambers, are generally placed 100 metre or more from the point of production of the π-mesons. But for time dilatation, very few π-mesons would ever reach such distant detectors.

One has another example of time dilatation with cosmic ray μ-mesons. The primary cosmic ray protons and α-particles produce π-mesons in high energy nuclear reactions near the top of the earth's atmosphere at altitudes of about 60 kilometre. These π-mesons decay quickly into μ-mesons. The mean lifetime of a μ-meson is $2 \cdot 2 \times 10^{-6}$ second, in the reference frame in which the μ-meson is at rest. Even if they travelled at the speed of light, if there were no time dilatation, the μ-mesons would only travel a mean distance of $3 \times 10^8 \times 2 \cdot 2 \times 10^{-6}$ or 660 metre before decaying, whereas in practice a substantial proportion of the μ-mesons in the cosmic rays do reach sea level. The mean energy of cosmic ray μ-mesons is about 3 GeV. Since $m_0 c^2 \sim 100$ MeV for μ-mesons, and from equation (2.29) $E = mc^2 = \gamma m_0 c^2$ we find

77

$\gamma \sim 30$. Allowing for time dilatation, the μ-mesons should travel a mean distance $\gamma v T_0 \sim 30 \times 0 \cdot 9995c \times 2 \cdot 2 \times 10^{-6}$ metre, that is $\sim 30 \times 660$ metre or ~ 20 kilometre before decaying, and a substantial proportion should reach sea level before decaying, in agreement with experiments. High energy nuclear physicists continually use the Lorentz transformations in their calculations. They can do so with complete confidence, since predictions based on the Lorentz transformations are invariably in agreement with experiments performed in the laboratory.

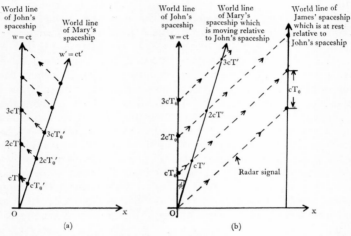

Figure 4.12. (a) Mary transmits radio signals at times 0, T_0', $2T_0'$, $3T_0'$, etc. on her clock. These signals reach John's spaceship at times 0, T, $2T$, $3T$, etc. on John's clock, where $T = K T_0'$. (b) John transmits radio signals from his spaceship at times 0, T_0, $2T_0$, etc. on his clock. The world lines of these signals are shown dotted. Mary is in a spaceship moving away from John's spaceship with uniform velocity v. John's radio signals reach Mary's spaceship at times 0, T', $2T'$, etc. on Mary's clock. James' spaceship is at rest relative to John's spaceship. James receives John's radio signals at time intervals of T_0 on his clock.

4.7 The Doppler effect*

In 1842 Doppler suggested that the motion of a light source affects the wavelength and frequency of the light. The K-calculus method is really an example of the Doppler effect. Consider the example shown previously in fig. 4.4 b and shown again in fig. 4.12 a. Mary's spaceship passes John's spaceship at $t = t' = 0$ and moves away with uniform velocity v relative to John's spaceship. Let Mary transmit radio signals at times $t' = 0$, T_0', $2T_0'$, etc. on her clock. Mary measures the frequency of the transmission of these signals from her spaceship to be $\nu_0 = 1/T_0'$. Let these signals reach John's spaceship

78

how can we say what is going on on the sun at this very moment? If the sun had stopped shining 5 minutes ago, it would take another 3 minutes for us to realize it, as there would still be some light on the way to the earth from the sun. To see what is happening on the far side of the moon we would have to send an astronaut or at least some equipment there to find out. Thus we must wait for information to be sent from distant events. Even if you look at your wrist watch, it takes a finite time for the light from your wrist watch to reach your eyes, and in principle you should allow for this propagation time. In practice this time would be $\sim 5 \times 10^{-9}$ second and can be ignored. However, if a television picture of Big Ben were transmitted to an astronaut on the moon it would take 1.28 second to reach him and he would have to allow for the propagation time of the signal.

Common-sense changes from generation to generation. In the Middle Ages, on the basis of limited geographical experiences, it was believed that the earth was flat. Children nowadays accept the roundness of the earth without question. We are beginning to appreciate the importance of propagation times. It is now possible to have radio and television programmes linking astronauts on the surface of the moon to the earth. The time delay due to propagation times is very noticeable when conversations take place between the astronauts and mission control on the earth. It takes about 1.28 second for the radio signals to reach the astronauts on the moon, and their replies take the same time to come back to the earth, so that there is a delay of about 3 second between the end of the transmission of a question from mission control and the receipt of the reply on earth.

As the range of our experiences widens there will be refinements of our ' common sense '. High energy nuclear physicists, who work with high energy particles most of the time, think in terms of the Lorentz transformations, and are so familiar with their successes that they would say that anybody who disputed the Lorentz transformations had no common sense at all. In principle the theory of special relativity is always better than Newtonian mechanics, but in the mechanical phenomena in our normal daily lives it is an unnecessary over-elaboration to use special relativity, since Newtonian mechanics and the concept of absolute time are satisfactory *approximations*. Special relativity is important in electricity and optics, and absolutely essential in high energy nuclear physics.

Absolute time was introduced into Newtonian mechanics as an assumption. The development of the theory of special relativity led people to question more carefully the experimental evidence in favour of the assumptions introduced into theories, and wherever possible to relate the quantities used in theories to how they can be measured experimentally. Theoretical quantities are still introduced into theories, for example, the electric field intensity. However, such quantities lead to predictions which can be tested by experiment. For example, the electric field intensity at a point can be used to calculate the force on a stationary test charge placed at that point.

It is often asked whether length contraction is real. What the principle of relativity says is that the laws of physics are the same in all inertial reference frames, but the actual measures of particular quantities may be different relative to different inertial reference frames. For example, if a ball rolls on the deck of a ship, which is moving with uniform velocity relative to the earth, the speed of the ball relative to the ship is different to the speed of the ball relative to the earth. Is this change real? According to the theory of special relativity not only the velocity of the ball relative to the ship and the earth will be different, but the measures of the length of the ship, the times of events, etc. will also be different relative to the two inertial reference frames. The laws of physics are the same in both systems, but the numerical values of physical quantities are different relative to the two systems.

It has been assumed throughout this chapter that Σ' (Mary's spaceship) always moves with *uniform* velocity v relative to Σ (John's spaceship). The theory of special relativity is only applicable to inertial reference frames. A brief discussion of accelerating and rotating non-inertial reference frames will be given in Chapter 6 and the behaviour of accelerating clocks will be considered in Chapter 7.

4.9 *Visual appearance of a rapidly moving object**

One should always say that a moving body is *measured* to be Lorentz contracted. What one sees (or photographs) depends on the light actually reaching the eye (or camera) at that given instant of time. For example, if one looks at the light coming from the stars in the heavens, some of the light will have left some of the stars millions of years ago, whereas the light from the moon left $1 \cdot 28$ second ago. Consider a cube of side l, which is moving with uniform velocity v comparable to c, relative to Σ, as shown in fig. 4.13 *a*, and in plan view in fig. 4.13 *b*. Let the cube be viewed from a *very large distance* in the symmetrical position shown in fig. 4.13 *a* by the observer who is at rest at O, the origin of Σ. Consider the light emitted from the corners A, B, C and D of the cube, when these corners are all at the same distance (measured relative to Σ) from O, the point of observation. The light from A, B, C and D reaches O at the same time. According to the theory of special relativity, the measured length of the sides AD and BC of the cube, which are perpendicular to v, are unchanged and equal to l, whereas AB and DC should be measured to be Lorentz contracted to $l\sqrt{(1-v^2/c^2)}$ relative to Σ. Hence the appearance of the face ABCD should be as shown in fig. 4.13 *c*. When the cube is moving relative to the observer, light from the corners E and F of the cube can also reach the eye at the same time as the light from A, B, C and D. Since the light from the corners E and F has farther to go

82

than the light from A, B, C and D, the light must leave E and F at an earlier time when the corners E and F were at E′ and F′ respectively, as shown in figs 4.13 a and b. At first sight it might appear to some readers that light from the corner E when it was at E′ would be blocked out by the face ADFE of the cube. However, provided the cube is very far away from O, the magnitude of v, the velocity of the cube to the right ($+x$ direction) in fig. 4.13 b is much greater than the component in that direction of the velocity of the light going from E′ to O.

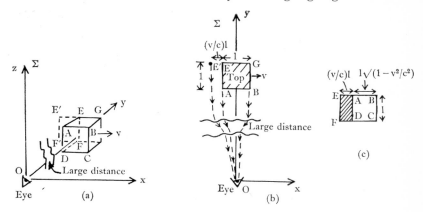

Figure 4.13. (a) The cube is moving with uniform velocity v relative to Σ. (b) Plan view of the moving cube. The light emitted from the corners A, B, C, D reaches the eye simultaneously. Light emitted from the corners E and F, when these corners are at the positions E′ and F′, reaches the eye at the same time as the light from A, B, C and D. The light from E′ travels an extra distance $\simeq l$ before reaching the eye taking an extra time l/c. In this extra time the cube moves a distance vl/c from E′ to E. (c) The relativistic picture of a distant cube.

The face ADFE of the moving cube should also be visible to the observer at O, as shown in fig. 4.13 c. If the observer at O is far away from the cube, the light from E′ and F′ must go an extra distance approximately equal to l, to reach O, as shown in fig. 4.13 b. The light from E′ takes a time l/c longer to reach the observer at O than the light from A. Since the cube moves at a speed v, EE′ must be equal to (vl/c), as shown in fig. 4.13 c. It will now be shown that this view of the moving cube has the same visual appearance as a stationary rotated cube. Let a stationary cube, of side l, be rotated through an angle α such that the projection of the side AE is equal to vl/c, as shown in fig. 4.14 a. The angle α is given by:

$$\sin \alpha = vl/c \div l = v/c.$$

The projection of the side AB is then equal to:

$$l \cos \alpha = l\sqrt{(1 - v^2/c^2)},$$

83

as shown in fig. 4.14 *b*. It can be seen that figs 4.13 *c* and 4.14*b* are the same, showing that a *distant* moving cube should *look* the same as a rotated stationary cube. There will be no distortion of perspective in this case due to the Lorentz contraction. (If the moving object is close to the eye, different parts of the object may appear rotated by different amounts leading to an apparent distortion of shape.)

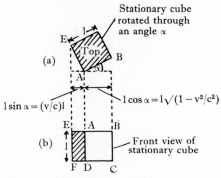

Figure 4.14. The visual appearance of a stationary cube rotated through an angle $\alpha = \sin^{-1}(v/c)$; the top view is shown in (*a*) and the front view in (*b*). The latter is the same as the visual appearance of a distant moving cube, predicted by the theory of special relativity, and illustrated in fig. 4.13 *c*.

4.10 *Intervals between events**

Time intervals and lengths of moving bodies were assumed to be invariants in Newtonian mechanics. This is not true in special relativity. However, a certain combination of the time interval between two events and the spatial separation of the two events is an invariant.

Consider two events, one at x, y, z at a time t and the other at $x + \delta x$, $y + \delta y$, $z + \delta z$ at a time $t + \delta t$ relative to the inertial reference frame Σ. From the Lorentz transformations, in an inertial frame Σ' moving with uniform velocity v relative to Σ:

$$x' = \gamma(x - vt); \quad y' = y; \quad z' = z; \quad t' = \gamma(t - vx/c^2);$$

and
$$x' + \delta x' = \gamma[x + \delta x - v(t + \delta t)]; \quad y' + \delta y' = y + \delta y;$$

$$z' + \delta z' = z + \delta z; \quad t' + \delta t' = \gamma[t + \delta t - v(x + \delta x)/c^2].$$

Subtracting: $\delta x' = \gamma(\delta x - v\delta t); \quad \delta y' = \delta y; \quad \delta z' = \delta z; \quad \delta t' = \gamma(\delta t - v\delta x/c^2)$. Notice the Lorentz transformations can be used to transform the increments δx, δy, δz and δt. Consider

$$\delta x'^2 + \delta y'^2 + \delta z'^2 - c^2\delta t'^2$$

$$= \gamma^2(\delta x - v\delta t)^2 + \delta y^2 + \delta z^2 - c^2\gamma^2(\delta t - v\delta x/c^2)^2$$

$$= \gamma^2[\delta x^2(1 - v^2/c^2) - c^2\delta t^2(1 - v^2/c^2)] + \delta y^2 + \delta z^2.$$

84

Since

$$\gamma = 1/\sqrt{(1 - v^2/c^2)},$$
$$\gamma^2 = 1/(1 - v^2/c^2).$$

Hence:

$$\delta x'^2 + \delta y'^2 + \delta z'^2 - c^2 \delta t'^2 = \delta x^2 + \delta y^2 + \delta z^2 - c^2 \delta t^2,$$

so that

$$\begin{aligned}\delta s^2 &= \delta x^2 + \delta y^2 + \delta z^2 - c^2 \delta t^2 \\ &= \delta x^2 + \delta y^2 + \delta z^2 - \delta w^2 \end{aligned} \tag{4.59}$$

is an invariant, that is, has the same numerical value in both Σ and Σ'. The quantity δs is called *the* interval between the events. According to the theory of special relativity, if two events occur, two observers, one at rest in Σ and one at rest in Σ', will record different distance and time separations between the two events, but they will record the same *interval*. If δs^2 is an invariant, then δs^2 cannot change sign when one transforms from one inertial reference frame to another.

If δs^2 is positive, then the interval δs between the two events is called a *space-like* interval and, if δs^2 is negative, δs is called a *time-like* interval.

To simplify the discussion in the rest of this section it will be assumed that δy and δz are always zero.

Consider the equation:

$$\delta x' = \gamma(\delta x - v\delta t).$$

The distance between the two events is zero in Σ', that is $\delta x' = 0$ if

$$\delta x = v\delta t \quad \text{or} \quad \frac{v}{c} = \frac{\delta x}{c\delta t}. \tag{4.60}$$

If δs is time-like, δs^2 is negative, $c^2 \delta t^2 > \delta x^2$ and $c|\delta t| > |\delta x|$ and v/c is less than unity in equation (4.60). Thus if δs is a time-like interval, it is always possible to find a reference frame in which $\delta x'$ is zero, that is, find a reference frame in which the two events are measured to occur at the same point of space. In this reference frame the time interval between the two events is the proper time interval between the events. For example, let one event occur at $x = 0$, $t = 0$ and the other at $x = 3 \times (3 \times 10^8)$ metre $= 3c$, $t = 5$ second (or $w = ct = 5c$), relative to the inertial reference frame Σ. From equation (4.59):

$$\delta s^2 = \delta x^2 - c^2 \delta t^2 = \delta x^2 - \delta w^2 = 9c^2 - 25c^2 = -16c^2.$$

Hence δs is a time-like interval. Substituting in equation (4.60) shows that $\delta x'$ is zero in a reference frame moving with velocity $v = 3c/5$

relative to Σ. In this reference frame, the proper time interval between the two events is:

$$\delta t' = \gamma(\delta t - v\delta x/c^2) = 1 \cdot 25 \ [5 - (3c/5)3c/c^2] = 4 \ \text{second}.$$

If δs were space-like $\delta x^2 > c^2\delta t^2$, $|\delta x| > c|\delta t|$, and in equation (4.60) v/c would have to be greater than unity. Hence it is impossible to find a reference frame in which $\delta x'$ is zero if the interval δs is space-like.

In the equation

$$\delta t' = \gamma\left(\delta t - \frac{v\delta x}{c^2}\right) = \gamma\delta t\left(1 - \frac{v\delta x}{c^2\delta t}\right) \qquad (4.61)$$

the time interval $\delta t'$ is zero if

$$\delta t = \frac{v\delta x}{c^2} \quad \text{or} \quad \frac{v}{c} = \frac{c\delta t}{\delta x}. \qquad (4.62)$$

If δs is space-like, δs^2 is positive, $\delta x^2 > c^2\delta t^2$ and $|\delta x| > c|\delta t|$. Hence, if δs is space-like, it is possible to satisfy equation (4.62) with v less than c and find an inertial reference frame in which $\delta t'$ is zero, that is, a reference frame in which the events are simultaneous. For example, if one event occurs at $x = 0$, $t = 0$ and the other at $x = 5c$, $t = 3$ second (or $w = ct = 3c$) relative to Σ, then

$$\delta s^2 = \delta x^2 - c^2\delta t^2 = (5c)^2 - c^2 3^2 = +16c^2,$$

so that δs is space-like. If in equation (4.61) we put $x = 5c$, $t = 3$ and $v = 3c/5$, we find that $\delta t'$ equals zero. If $v/c > c|\delta t|/|\delta x|$, but v/c remains less than unity, then $\delta t'$ is opposite in sign to δt, and the time order of the two events is reversed. For example, if for $\delta t = 3$, $\delta x = 5c$ we put $v = 4c/5$ in equation (4.61) we find that $\delta t' = -5/3$ second, so that in a reference frame Σ' moving with velocity $4c/5$ relative to Σ, the event which is at $x = 0$, $t = 0$ relative to Σ is measured to occur $1\frac{2}{3}$ second after the event which is at $x = 5c$, $t = 3$ relative to Σ.

At first sight it seems to go against our common sense to find that the measured time order of events can be different in different inertial reference frames. This reversal of the time order of events can only occur if δs^2 is space-like, that is, if $\delta x^2 > c^2\delta t^2$ or $|\delta x| > c|\delta t|$. If $|\delta x| > c|\delta t|$, it is impossible to send a light signal the distance $|\delta x|$ between the events in the time interval $|\delta t|$. Since energy and momentum cannot be transmitted at a speed exceeding the speed of light, if δs is positive, what happens at one event cannot influence what happens at the other event. Since, when δs^2 is positive, one event cannot influence what happens at the other event, it does not matter which event is measured to occur first. One cannot influence what is going on at present, or in the next few years, on a star a thousand light years away, since, if we sent a radio signal now, it would not reach the star for a thousand years. One cannot influence what is going on on the moon in one second's time, since a radio signal would take $1\cdot28$ second to reach the moon. When δs^2 is time-like, then it is possible to send a light or radio signal from one event to the other. For example, we could influence what happens on the moon in a month's time by sending a spaceship to get there in time. Since $v<c$, when δs^2 is negative and $|\delta x|<c\,|\delta t\,|$, then in equation (4.61) $v\delta x/c^2\delta t<1$ and $\delta t'$ always has the same sign as δt, so that, when δs^2 is time-like and causal connection between the two events is possible, there is a definite time order for the two events which is the same in all inertial reference frames. These conclusions can be illustrated graphically.

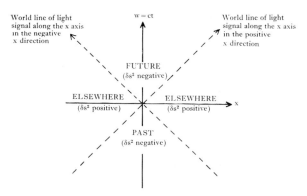

Figure 4.15. One event takes place at the origin at $x=0$, $t=0$. If another event takes place in either of the regions marked PAST or PRESENT, then δs^2 is negative, light signals can go from one of the events to the other and one of the events can influence what happens at the other event. If the other event is in the ELSEWHERE regions δs^2 is positive and one event cannot influence what happens at the other.

It will be assumed that one event is here and now at the point $x=0$ at the time $t=0$ at the origin of Σ. A space–time diagram will be used to plot the positions and times of other events, as shown in fig. 4.15. The world lines of light rays passing the origin $x=0$ at $t=0$ are given by $x=\pm ct$ and are at $45°$ to the x and $w=ct$ axes, as shown in fig. 4.15. Events in the region labelled FUTURE in fig. 4.15 can be reached from the origin with speeds less than the speed of light, so that the reader can influence all events in this part of fig. 4.15. His future world line is in this region also. Events in the region labelled PAST could have sent signals to reach the reader at the origin at or before $t=0$ and could influence what happens at $x=0$, $t=0$. The world line of the reader's past history is in the PAST region. Events in the regions labelled ELSEWHERE cannot send light signals to reach $x=0$ by $t=0$, neither can light signals be sent from the origin at $x = 0$, $t = 0$ to reach these events to have any influence on them. The intervals between the event at the origin, $x = 0$ at $t = 0$, and events in either the PAST or FUTURE regions are time-like ($\delta s^2 < 0$), and the intervals between the event at the origin and events in the ELSEWHERE regions are space-like ($\delta s^2 > 0$).

4.11 *Minkowski diagrams**

In this chapter, in particular in §§ 4.2, 4.3 and 4.4, when space–time diagrams were drawn, a rectangular co-ordinate system with axes Ox and Ow, where $w = ct$, was used to plot the positions and times of events relative to Σ, the inertial reference frame in which John's spaceship was at rest at the origin. It will now be shown how the

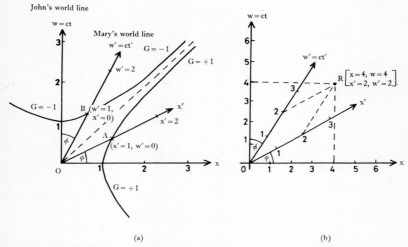

(a)   (b)

Figure 4.16.   (*a*) A Minkowski diagram.   The rectangular axes Ox, Ow are used to represent the co-ordinates and times of events relative to Σ, the inertial reference frame in which John's spaceship is at rest at the origin.   The oblique axes Ox', Ow' can be used to represent the positions of the events relative to the inertial frame Σ' in which Mary's spaceship is at rest at the origin.   (*b*) A Minkowski diagram for the case when $v = 0 \cdot 6c$.   The event $x = 4$, $w = 4$ is shown.   Relative to the oblique axes Ox', Ow', this event has co-ordinates $x' = 2$, $w' = 2$.

same world points and world lines can be used to represent the positions and times of the events relative to the inertial reference frame Σ', which moves with uniform velocity v relative to Σ, and in which Mary's spaceship is at rest at the origin, provided oblique axes are used to represent the positions and times of events relative to Σ'.

In fig. 4.16 *a* rectangular axes Ox, Ow are used to represent the positions and times of events relative to Σ.   On the Ow', or time axis, of the oblique axes which will be used to represent the positions

88

and times of events relative to Σ', we must have $x'=0$. From the Lorentz transformations, if $x'=0$:

$$x' = \gamma(x - vt) = 0$$

or

$$x = vt = \left(\frac{v}{c}\right)ct = \left(\frac{v}{c}\right)w. \qquad (4.63)$$

The line $x=(v/c)w$ is represented by Ow' in fig. 4.16 a. It is at an angle $\phi = \tan^{-1}(v/c)$ to the Ow axis in fig. 4.16 a. This is the world line of Mary's spaceship relative to the Ox, Ow rectangular axes.

Along the Ox' axis of the oblique co-ordinate system which will be used to represent the positions and times of events relative to Mary's spaceship (Σ'), $w' = ct' = 0$. From the Lorentz transformations, if $t' = 0$:

$$t' = \gamma(t - vx/c^2) = 0$$

or

$$w = ct = (v/c)x.$$

The line $w'=0$ is represented by Ox' in fig. 4.16 a. It makes an angle ϕ with the Ox axis, where $\phi = \tan^{-1}(v/c)$, as shown in fig. 4.16 a. The lines Ox' and Ow' will now be used as oblique axes to represent the co-ordinates and times of events relative to Σ'.

The unit of length along the Ox' axis is the distance from the origin O to the point A at $x'=1$, $w'=ct'=0$ in fig. 4.16 a. Relative to the rectangular axes Ox, Ow, according to the Lorentz transformations, the point A has co-ordinates:

$$x = \gamma(x' + vt') = \gamma x' = \gamma,$$

$$w = ct = \gamma(ct' + vx'/c) = \gamma v/c.$$

Using the scales appropriate to the rectangular axes Ox, Ow the distance of the point A from the origin is:

$$g_1 = \sqrt{(\gamma^2 + \gamma^2 v^2/c^2)} = \gamma\sqrt{(1 + v^2/c^2)} = \sqrt{\left(\frac{1 + v^2/c^2}{1 - v^2/c^2}\right)}. \qquad (4.64)$$

Similarly the unit of w' along Ow' is the distance from the origin to the point B at $w' = ct' = 1$, $x' = 0$ in fig. 4.16 a. Using the Lorentz transformations, for the point B:

$$x = \gamma(x' + vt') = \gamma[x' + (v/c)w'] = \gamma v/c,$$

$$w = ct = \gamma(ct' + vx'/c) = \gamma.$$

Using the scales appropriate to the Ox, Ow rectangular axes, the distance from O to B is given by:

$$g_2 = \sqrt{(\gamma^2 + \gamma^2 v^2/c^2)} = \gamma\sqrt{(1 + v^2/c^2)} = \sqrt{\left(\frac{1 + v^2/c^2}{1 - v^2/c^2}\right)}. \quad (4.65)$$

Now if we use Ox' and Ow' in fig. 4.16 a as oblique axes and change our scales of x' and w' according to equations (4.64) and (4.65), then one can use the *same* world points and world lines to represent the positions and times of events relative to Σ', using Ox' and Ow' as oblique axes, and relative to Σ, using the rectangular Ox, Ow axes. A proof is given by Rosser[1 b]. We shall merely use a simple numerical example to illustrate the invariance of world points.

The reader should make a large-scale drawing of fig. 4.16 b. Start by drawing rectangular axes Ox, Ow to represent the positions and times of events relative to Σ. Let Σ' move with velocity $v = 3c/5$ relative to Σ. Hence $\phi = \tan^{-1}\frac{3}{5}$. Draw the Ox' and Ow' axes. For example, join the origin to the point $x = 5$, $w = 3$ to get the line Ox'. If $v/c = \frac{3}{5}$:

$$g_1 = g_2 = \sqrt{\left(\frac{1 + v^2/c^2}{1 - v^2/c^2}\right)} = \sqrt{\left(\frac{1 + 3^2/5^2}{1 - 3^2/5^2}\right)} = \sqrt{\left(\frac{34}{16}\right)} = 1 \cdot 46.$$

Mark out units of $x' = 1$, 2, 3, 4, etc. along Ox', using the distance from O to $x' = 1$ as $1 \cdot 46$ on the scale used originally for the rectangular axes Ox, Ow. For example, if the unit of x is 1 cm long and the unit of $w = ct$ is 1 cm long in the Ox, Ow rectangular axes, then the distance from O to $x' = 1$ along Ox' is $1 \cdot 46$ cm. Similarly, mark out $w' = 1$, 2, 3, 4, etc. on the Ow' axis, making the distance from O to $w' = 1$ equal to $1 \cdot 46$ times the length from O to $w = 1$. Plot the world point R at $x = 4$, $w = ct = 4$ relative to Σ, using the Ox, Ow axes, as shown in fig. 4.16 b. To determine the co-ordinates of this world point relative to the Ox', Ow' oblique axes, draw lines through the world point R parallel to the Ow' and Ox' axes, as shown in fig. 4.16 b. It can be seen from the graph that these lines cut the oblique axes at $x' = 2$ and $w' = 2$ respectively. If the position of the world point R need not be changed, we would expect the co-ordinates of the event at $x = 4$, $w = 4$ relative to Σ to be at $x' = 2$, $w' = 2$ relative to Σ'. For $v = 3c/5$:

$$\gamma = 1/\sqrt{(1 - 3^2/5^2)} = 1/\sqrt{(1 - 9/25)} = 1/\sqrt{(16/25)} = 5/4.$$

Using the Lorentz transformations, for $x = 4$, $w = ct = 4$:

$$x' = \gamma\left(x - \frac{v}{c}ct\right) = \frac{5}{4}\left(4 - \frac{3}{5} \times 4\right) = 2,$$

$$t' = \gamma(t - vx/c^2),$$

$$w' = ct' = \gamma\left(ct - \frac{v}{c}x\right) = \frac{5}{4}\left(4 - \frac{3}{5} \times 4\right) = 2.$$

This shows that, with the use of the oblique axes Ox' and Ow', and the choice of lengths of units of x' and w' given by equations (4.64) and (4.65) respectively, the positions of world points and world lines need not be changed in fig. 4.16 b, and the *same* world points and world lines can be used to represent the positions and times of events relative to Σ, using the Ox, Ow axes and relative to Σ' using the oblique Ox', Ow' axes. This is true whatever the value of v, provided it is less than c. Thus the *same* world points and world lines can be used to represent the positions of events relative to all inertial reference frames moving along the x axis of Σ, and whose origins coincide with the origin of Σ at $t = t' = 0$. Thus John and Mary can use the same world points on the same graph, if John uses the rectangular axes Ox, Ow and Mary uses the oblique axes Ox', Ow', provided Mary uses different scales to John for x' and w', namely the scales given by equations (4.64) and (4.65). (We could have used rectangular axes to represent the positions and times of events relative to the inertial reference frame Σ' in which Mary's spaceship is at rest, but we would then have had to use oblique axes to represent the world points relative to the reference frame Σ in which John's spaceship is at rest.)

For the co-ordinate system Ox', Ow' in fig. 4.16 a, used to represent the positions and times of events relative to the inertial reference frame Σ' in which Mary's spaceship is at rest, the length of the unit of x' along the Ox' axis is given by the length OA from the origin to the point A which has co-ordinates $x' = 1$, $w' = 0$. According to the Lorentz transformations, relative to the rectangular co-ordinate system Ox, Ow used to represent the positions and times of events relative to Σ, the inertial reference frame in which John's spaceship is at rest, the point A has the co-ordinates $x = \gamma$, $w = ct = \gamma v/c$. For the point A:

$$x^2 - w^2 = x^2 - c^2 t^2 = \gamma^2 - \gamma^2 v^2/c^2 = \gamma^2(1 - v^2/c^2) = +1.$$

The expression $(x^2 - w^2)$ is independent of v, the velocity of Σ' relative to Σ. Relative to the rectangular co-ordinate system Ox, Ow, the point A, whose distance from the origin gives the length of the unit of x' for the oblique axes Ox', Ow' used to represent the positions of world points relative to Σ', always lies on the curve:

$$x^2 - w^2 = x^2 - c^2 t^2 = G = +1, \tag{4.66}$$

whatever the value of v. If the velocity of Σ' relative to Σ were different, the position of A, which has co-ordinates $x = \gamma$, $w = \gamma v/c$ would vary since v varies, but A would always be on the curve $(x^2 - w^2) = +1$. Equation (4.66) is the equation of a hyperbola. It is shown in fig. 4.16 a. The Ox' axis appropriate to any inertial reference frame Σ' moving with uniform velocity v along the x axis of Σ, and whose origin coincides with the origin of Σ at $t = t' = 0$, is at an angle $\phi = \tan^{-1}(v/c)$ to the Ox axis. The unit of length for

91

this Ox' axis is the distance from the origin to the point of intersection of this Ox' axis with the hyperbola given by equation (4.66) corresponding to $G = +1$.

The unit of w' in fig. 4.16 a is given by the length OB, where B corresponds to the point $x' = 0$, $w' = ct' = 1$. According to the Lorentz transformations the x and w co-ordinates of B are $\gamma v/c$ and γ respectively. For the point B, relative to the rectangular co-ordinate system Ox, Ow, we have:

$$x^2 - w^2 = x^2 - c^2 t^2 = \gamma^2 v^2/c^2 - \gamma^2 = -1.$$

Thus, whatever the value of v, B always lies on the hyperbola:

$$x^2 - w^2 = G = -1. \tag{4.67}$$

The Ow' axis, appropriate to the inertial reference frame Σ' which is moving with uniform velocity v relative to Σ, is at an angle $(\pi/2 - \phi)$ to the Ox axis, where $\phi = \tan^{-1}(v/c)$, as shown in fig. 4.16 a. The unit of w' in the oblique co-ordinate system Ox', Ow' is given by the distance from the origin to the point of intersection of the Ow' axis and the hyperbola given by equation (4.67) corresponding to $G = -1$. The hyperbolae given by equations (4.66) and (4.67) are known as the calibration hyperbolae. Diagrams such as fig. 4.16 a are known as *Minkowski diagrams*, in honour of H. Minkowski who first developed the method in 1908.

If a particle moves in the xy plane of Σ, we would need three dimensions x, y, $w = ct$ to represent its displacement at various times. For motion in three dimensions, one would have to use four-dimensional space–time diagrams, using x, y, z, $w = ct$. Since $y' = y$ and $z' = z$, there is no need to change the units of y' and z' and the Oy and Oy' and the Oz and Oz' axes are coincident. Minkowski suggested that the four-dimensional space x, y, z, $w = ct$ represents 'the world'. In this world, the world points and world lines are invariants.

Minkowski diagrams can be used to illustrate non-absolute simultaneity, time dilatation and length contraction. We shall use the same example as shown in fig. 4.16 b, and shown again on an enlarged scale in figs. 4.17 a and 4.17 b. The reader should make his own large-scale diagrams to check our conclusions. It will be recalled that Mary's spaceship (Σ') moves with uniform velocity $v = 0 \cdot 6c$ relative to John's spaceship (Σ), such that $\tan \phi = 0 \cdot 6$, $\gamma = 5/4$ and $g_1 = g_2 = 1 \cdot 46$. (Note the different lengths of the units of x and x' and of w and w' in figs 4.17 a and 4.17 b consistent with $g_1 = g_2 = 1 \cdot 46$.)

Consider two events, one at the origin O at $x' = 0$, $w' = ct' = 0$ relative to Σ' and the other at world point P in fig. 4.17 a at $x' = 1$, $w' = ct' = 0$ on the Ox' axis of the oblique axes (Σ'). These events are simultaneous at $t' = 0$ relative to Σ'. The event at the origin O is at $w = ct = 0$ relative to Σ. In order to determine the time of the event at the world point P relative to the rectangular axes (Σ), draw a line through P parallel to the Ox axis to cut the Ow axis at the world point Q. From fig. 4.17 a it can be seen that at Q, graphically $w = ct = 0 \cdot 75$. Thus relative to Σ the event at world point P does not occur at the same time as the event at O at the origin, though the events at O and P are simultaneous at $t' = 0$, relative to Σ'. From the Lorentz transformations for $x' = 1$, $t' = 0$:

$$t = \gamma(t' + vx'/c^2) = \gamma v/c^2,$$

$$w = ct = \gamma v/c = \tfrac{5}{4} \times \tfrac{3}{5} = 0 \cdot 75.$$

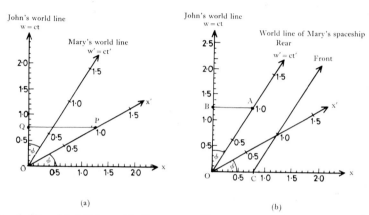

(a)

(b)

Figure 4.17. (a) Minkowski diagram to illustrate non-absolute simultaneity. (b) Minkowski diagram to illustrate time dilatation and length contraction.

Thus the graphical method using a Minkowski diagram gives the same result as the Lorentz transformations. It is left as an exercise for the reader to show that events at $x = 0$ and $x = 1$ at the time $w = ct = 0$ relative to Σ are not simultaneous relative to Σ'.

To illustrate time dilatation consider Event 1 at the origin O in fig. 4.17 b at $x' = 0$, $w' = ct' = 0$ relative to the oblique axes (Σ') and Event 2 at world point A at $x' = 0$, $w' = ct' = 1$ relative to Σ'. The time of the event at the origin O is $w = ct = 0$ relative to the rectangular axes (Σ). To determine the time of the Event 2 at world point A, relative to the rectangular axes (Σ), draw a line through A parallel to the Ox axis to cut the Ow axis at world point B. It can be seen from fig. 4.17 b that at B, $w = ct = 1 \cdot 25$. Thus the time between the two events at world points O and A is $\delta w = c\delta t = 1 \cdot 25$ relative to Σ and $\delta w' = c\delta t' = 1$ in the inertial reference frame Σ', in which the events occur at the same point $x' = 0$, illustrating time dilatation. From the Lorentz transformations, for $x' = 0$, $w' = ct' = 1$:

$$w = ct = \gamma(ct' + vx'/c) = \gamma ct'.$$

93

Therefore,

$$\delta w = c\delta t = \gamma c\delta t' = \tfrac{5}{4} \times 1 = \tfrac{5}{4}.$$

Thus the graphical method using a Minkowski diagram and the Lorentz transformations give the same result.

As an example of length contraction, assume that Mary's spaceship is of unit length, that the rear of Mary's spaceship is at $x' = 0$ and the front is at $x' = 1$ in Σ'. The world lines of the front and rear of Mary's spaceship are shown in fig. 4.17 b. To measure the length of Mary's spaceship relative to Σ, John must measure the distance between the world lines of the front and rear of Mary's spaceship at the same time relative to Σ, say at $t = 0$. Thus, relative to the inertial reference frame Σ in which John's spaceship is at rest, the length of Mary's spaceship, measured at the time $t = 0$ in Σ, is equal to the distance OC between the world lines of the front and rear of Mary's spaceship, which is equal to $0 \cdot 8$ in fig. 4.17 b. According to the Lorentz transformations if $x' = 1$ in Σ', at $t = 0$ in Σ we have:

$$x' = \gamma(x - vt) = \gamma x,$$

$$x = x'/\gamma = 1/1 \cdot 25 = 0 \cdot 8.$$

Thus the graphical method using a Minkowski diagram again gives the same result as the Lorentz transformations.

It is left as an exercise for the reader to show graphically that time dilatation and length contraction are reciprocal, by showing that the event at $x = 0$, $w = ct = 1$ in Σ is at $w' = ct' = 1 \cdot 25$ relative to Σ', and that if John's spaceship is of unit length in Σ, then relative to Σ' its length is $0 \cdot 8$. A reader interested in a fuller account of Minkowski diagrams and an account of the development of special relativity using four vectors is referred to Rosser[1 b].

References

1. ROSSER, W. G. V., *Introductory Relativity* (Butterworths, London, 1967). (a) p. 99 and 162; (b) Ch. 6.

Problems

(Assume that the velocity of light $c = 3 \cdot 00 \times 10^8$ m s^{-1})

4.1. Assume that the inertial reference frame Σ' moves with uniform velocity $0 \cdot 6c$ relative to Σ along the x axis of Σ. Let the origins of Σ and Σ' coincide at $t = t' = 0$. Use the Lorentz transformations to:
    (a) Find the co-ordinates and times relative to Σ' of the events, which have the following co-ordinates and times relative to Σ:
        (i) $x = 4$ m; $t = 6$ s;
        (ii) $x = 7 \times 10^8$ m; $t = 2$ s;
        (iii) $x = 6 \times 10^{10}$ m; $t = 3$ s.

(b) Find the co-ordinates and times relative to Σ of the events, which have the following co-ordinates and times relative to Σ':
(i) $x' = 10$ m; $t' = 4$ s;
(ii) $x' = 9 \times 10^9$ m; $t' = 4$ s;
(iii) $x' = 10^{11}$ m; $t' = 50$ s.

4.2. By what amount is the earth shortened along its diameter (as measured by an observer at rest relative to the sun) due to the orbital motion of the earth around the sun? (Take the velocity of the earth as 30 kilometre per second and the radius of the earth as 6371 kilometre.)

4.3. A spaceship is moving at such a speed in the laboratory system that its measured length is half its proper length. How fast is the spaceship moving relative to the laboratory system?

4.4. If the mean lifetime of a μ-meson when it is at rest is $2 \cdot 2 \times 10^{-6}$ s, calculate the average distance it will travel *in vacuo* before decay, if its velocity is (a) $0 \cdot 9c$; (b) $0 \cdot 99c$; (c) $0 \cdot 999c$. (Hint: Use equation 4·54.)

4.5. A physicist tells a friend, whose mass is 150 kg, that the best way to slim is to move so fast relative to the laboratory that he is Lorentz-contracted.
(a) What speed would he have to move to ' reduce ' his measured dimensions in the direction of motion to half his laboratory size ?
(b) What would be his mass relative to the laboratory?

4.6. Calculate the wavelength (using the relativistic formula) for light of wavelength 500 nm when the source is approaching the observer with velocity (a) $0 \cdot 1c$, (b) $0 \cdot 9c$.

4.7. A physicist is arrested for driving through the red lights at a road junction. At the trial the physicist claims he was driving so fast that the red light appeared green to him. ' Plea accepted ', said the judge, ' but I fine you a pound for each kilometre per hour your speed exceeded the speed limit of 45 kilometre per hour '. Calculate the fine, taking the wavelength of green light to be 530 nm and the wavelength of red light to be 630 nm.

4.8. One event occurs at the origin of an inertial frame Σ at the time $t=0$. Another event occurs at $x=3c$, $y=z=0$ at a time $t=6$ s relative to Σ.
(a) Determine the velocity (relative to Σ) of the inertial frame Σ' in which the two events are recorded at the same point in space, and
(b) what is the time interval between the events relative to Σ'?

4.9. (a) An event occurs at $x=3c$, $t=4$ s in the inertial reference frame Σ. Use a Minkowski diagram to determine the co-ordinates and time of the event relative to a reference frame Σ' moving with uniform velocity $0 \cdot 6c$ relative to Σ. Find the velocity relative to Σ of the reference frame in which the event occurs at $x'=0$.
(b) If an event is at $x=4c$, $t=3$ s in Σ, use a Minkowski diagram to find its co-ordinates relative to a reference frame moving with velocity $0 \cdot 6c$ relative to Σ. Find the velocity relative to Σ of the reference frame in which the event occurs at $t'=0$.
Check your answers, using the Lorentz transformations.

CHAPTER 5
the transformations of special relativity

5.1 Introduction

IN Chapter 4 we only discussed how the co-ordinates and times of events are transformed from one inertial reference frame to another. In practice one is often interested in how other quantities, such as the velocity, momentum and energy of a particle change, if one changes one's standard of rest, that is, if one transforms from one inertial reference frame to another moving relative to it. It will be assumed throughout this chapter that the inertial reference frame Σ' moves with uniform velocity v relative to Σ along their common x axis and that the origins of Σ and Σ' coincide at $t = t' = 0$. Some applications of the transformations will be discussed.

5.2 The velocity transformations

When a particle is moving relative to an inertial reference frame its velocity relative to that inertial reference frame is defined as the distance it moves per unit time, measured relative to that inertial reference frame. Let a particle be measured to be at a point x, y, z at a time t relative to the inertial frame Σ, and let it be measured to be at $x + \delta x$, $y + \delta y$, $z + \delta z$ at a time $t + \delta t$ relative to Σ. The velocity of the particle, relative to Σ, is defined as a vector $\mathbf{u}$ having components:

$$u_x = \frac{\delta x}{\delta t}; \quad u_y = \frac{\delta y}{\delta t}; \quad u_z = \frac{\delta z}{\delta t}. \tag{5.1}$$

The magnitude of u is given by:

$$u^2 = u_x{}^2 + u_y{}^2 + u_z{}^2. \tag{5.2}$$

The two successive measurements of the position of the particle are events, since they involve measurements such as the passing of the particle past marks on a scale, or the passage of a high energy atomic particle through two scintillation counters. The co-ordinates and times of these events relative to Σ' can be calculated using the Lorentz transformations. Corresponding to x, y, z, t in Σ one has in Σ':

$$x' = \gamma(x - vt); \quad y' = y; \quad z' = z; \quad t' = \gamma(t - vx/c^2), \tag{5.3}$$

and, corresponding to $x + \delta x$, $y + \delta y$, $z + \delta z$, $t + \delta t$, in Σ, we have in Σ':

$$\left. \begin{array}{l} x' + \delta x' = \gamma[x + \delta x - v(t + \delta t)]; \quad y' + \delta y' = y + \delta y, \\ z' + \delta z' = z + \delta z; \quad t' + \delta t' = \gamma[t + \delta t - v(x + \delta x)/c^2]. \end{array} \right\} \tag{5.4}$$

Subtracting (5.3) from (5.4), we have:

$$\delta x' = \gamma(\delta x - v\delta t); \quad \delta y' = \delta y; \quad \delta z' = \delta z; \quad \delta t' = \gamma(\delta t - v\delta x/c^2). \quad (5.5)$$

The velocity of the particle, measured in Σ', has components:

$$u_x' = \frac{\delta x'}{\delta t'} = \frac{\gamma(\delta x - v\delta t)}{\gamma(\delta t - v\delta x/c^2)} = \frac{(\delta x/\delta t - v)}{[1 - v(\delta x/\delta t)/c^2]} = \frac{(u_x - v)}{(1 - vu_x/c^2)}, \quad (5.6)$$

$$u_y' = \frac{\delta y'}{\delta t'} = \frac{\delta y}{\gamma(\delta t - v\delta x/c^2)} = \frac{(\delta y/\delta t)\sqrt{(1 - v^2/c^2)}}{[1 - v(\delta x/\delta t)/c^2]} = \frac{u_y\sqrt{(1 - v^2/c^2)}}{(1 - vu_x/c^2)},$$
$$(5.7)$$

$$u_z' = \frac{\delta z'}{\delta t'} = \frac{u_z\sqrt{(1 - v^2/c^2)}}{(1 - vu_x/c^2)}. \quad (5.8)$$

The inverse relations can be obtained by changing unprimed quantities into primed quantities, changing primed quantities into unprimed quantities and replacing v by $-v$, giving:

$$u_x = \frac{(u_x' + v)}{(1 + vu_x'/c^2)}; \quad u_y = \frac{u_y'\sqrt{(1 - v^2/c^2)}}{(1 + vu_x'/c^2)}; \quad u_z = \frac{u_z'\sqrt{(1 - v^2/c^2)}}{(1 + vu_x'/c^2)}. \quad (5.9)$$

Notice, if $u \ll c$ and $v \ll c$, from equations (5.6), (5.7) and (5.8) we have:

$$u_x' \to (u_x - v); \quad u_y' \to u_y; \quad u_z' \to u_z.$$

This is in agreement with the Galilean velocity transformations of Newtonian mechanics.

As a typical example from Newtonian mechanics, consider a ship moving with a uniform velocity of 18 metre per second relative to the earth. Let a ball be rolled at a speed of $u_x' = 2$ metre per second relative to the ship, in the direction of motion of the ship. From equations (5.9) the speed of the ball relative to the earth is:

$$u_x = \frac{(u_x' + v)}{(1 + vu_x'/c^2)} = \frac{(18 + 2)}{(1 + 2 \times 18/(3 \times 10^8)^2)} = \frac{20}{(1 + 4 \times 10^{-16})}.$$

Expanding the denominator using the binomial theorem, we have:

$$u_x \simeq 20(1 + 4 \times 10^{-16})^{-1} = 20 - 80 \times 10^{-16} = 20 - 8 \times 10^{-15}$$

$$= 19 \cdot 999\ 999\ 999\ 999\ 992 \text{ metre per second.}$$

According to the Galilean transformations, equation (1.18):

$$u_x = u_x' + v = 18 + 2 = 20 \text{ metre per second.}$$

This example illustrates how, in normal circumstances, the deviations from Newtonian mechanics are negligible.

4   97

Let a radioactive atom move with a velocity $v = 0 \cdot 1c$ along the x axis of the laboratory system Σ. Let it emit a β-particle of velocity $0 \cdot 95c$ relative to the inertial reference frame Σ' in which the radioactive atom is at rest. If the β-particle is emitted along the x' axis of Σ', such that $u_x' = 0 \cdot 95c$, its speed relative to the laboratory system Σ is:

$$u_x = \frac{(u_x' + v)}{(1 + v u_x'/c^2)} = \frac{0 \cdot 95c + 0 \cdot 1c}{1 + 0 \cdot 1 \times 0 \cdot 95} = \frac{1 \cdot 05c}{1 \cdot 095}$$
$$= 0 \cdot 959c.$$

According to the Galilean transformations:

$$u_x = u_x' + v = 0 \cdot 95c + 0 \cdot 1c = 1 \cdot 05c.$$

Thus the deviations from the Galilean transformations are important in high energy nuclear physics.

If $u_x' = 0 \cdot 9c$ and $v = 0 \cdot 9c$, then

$$u_x = \frac{0 \cdot 9c + 0 \cdot 9c}{1 + 0 \cdot 9 \times 0 \cdot 9} = \frac{1 \cdot 8c}{1 \cdot 81} = 0 \cdot 994\ c.$$

The reader can check that one cannot obtain velocities exceeding c by adding a number of velocities which are themselves less than c.

If $u_x' = c$ in Σ', then in Σ:

$$u_x = \frac{(u_x' + v)}{(1 + v u_x'/c^2)} = \frac{c + v}{1 + vc/c^2} = \frac{c + v}{1 + v/c} = \frac{c(c + v)}{(c + v)} = c.$$

Thus the speed of light in empty space has the same numerical value in Σ and Σ', illustrating that the velocity transformations are consistent with the principle of the constancy of the speed of light, as of course they should be, since the Lorentz transformations were developed from the principle of the constancy of the speed of light in Chapter 4.

5.3 *Fizeau's experiment*

In 1851 Fizeau showed that the speed of light in moving water depended on the speed of the water relative to the laboratory.

Let Σ be the laboratory system in which the water is moving with uniform velocity v in the positive x direction, as shown in fig. 5.1 *a*. Let the direction of the light be in the positive x direction also, as shown in fig. 5.1 *a*. In the inertial frame Σ' moving with uniform velocity v relative to Σ, the water is at rest, as shown in fig. 5.1 *b*. Let the refractive index of stationary water be n' measured in Σ'. The velocity of the light in Σ' has the components:

$$u_x' = c/n'; \quad u_y' = u_z' = 0.$$

Applying equations (5.9):

$$u_y = u_z = 0,$$

$$u_x = \frac{(u_x' + v)}{(1 + vu_x'/c^2)} = \frac{(c/n' + v)}{(1 + vc/n'c^2)} = \frac{c}{n'}\left[1 + \frac{n'v}{c}\right]\left[1 + \frac{v}{n'c}\right]^{-1}.$$

Expanding $(1 + v/n'c)^{-1}$ by the binomial theorem and neglecting terms of order v^2/c^2, we have:

$$u_x \simeq \frac{c}{n'}\left[1 + \frac{n'v}{c}\right]\left[1 - \frac{v}{n'c}\right],$$

$$u_x \simeq \frac{c}{n'} + v\left[1 - \frac{1}{n'^2}\right]. \qquad\qquad (5.10)$$

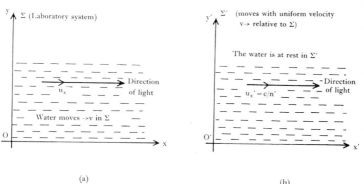

(a)　　　　　　　　　　(b)

Figure 5.1.　In Σ' the water is at rest, and the speed of light is the same and equal to c/n' in all directions in Σ', as shown in (b).　In Σ, the laboratory system, the water is moving with velocity v, as shown in (a).　According to the velocity transformations of special relativity, the speed of light parallel to the x axis is equal to $c/n' + v(1 - 1/n'^2)$ in Σ.

This value for the speed of light in moving water is in agreement with the results obtained by Fizeau.　*If* the Galilean transformations were correct, one would have:

$$u_x = u_x' + v = (c/n') + v. \qquad\qquad (5.11)$$

This value is *not* in agreement with the experimental results.　The reason the Galilean transformations are inaccurate in this case is that one of the velocities, namely c/n', is comparable to c, since $n' = 1 \cdot 33$. If n' were, say, 1000, then $1 - 1/n'^2$ would be $1 - 10^{-6}$ and equation (5.11) would be a satisfactory approximation to equation (5.10).

By measuring the speed of light in moving water it is possible to determine the speed of the water.　This does not contravene the principle of relativity.　If an astronaut, inside a spaceship moving with uniform velocity relative to the fixed stars, repeated Fizeau's experiment, all he could determine would be the speed of the water relative to his spaceship.　He could not determine the speed of the spaceship relative to the earth by this experiment.

5.4 *Other transformations of high energy physics*

The theory of high energy physics was developed from *experiments* in Chapter 2. These equations can be applied in the laboratory system. It is sometimes more convenient to carry out the calculations in an inertial reference frame moving relative to the laboratory, such as, for example, the centre of mass system. The appropriate transformations will now be developed. Some readers may find the algebra a little tedious. If they are of the believing type, they may prefer merely to accept the transformations derived in this section. The quantity $1/\sqrt{(1-u^2/c^2)}$ appears frequently, and we shall start by deriving the transformation for this quantity.

(*a*) *Transformation of* $\sqrt{(1-u^2/c^2)}$ *and* $\sqrt{(1-u'^2/c^2)}$

The total velocity of a particle relative to Σ' is given by:

$$u'^2 = u_x'^2 + u_y'^2 + u_z'^2.$$

Using equations (5.6), (5.7) and (5.8):

$$u'^2 = \frac{(u_x-v)^2 + (1-v^2/c^2)u_y^2 + (1-v^2/c^2)u_z^2}{(1-vu_x/c^2)^2}.$$

Remembering,

$$u^2 = u_x^2 + u_y^2 + u_z^2$$

or

$$u_y^2 + u_z^2 = u^2 - u_x^2,$$

we have:

$$u'^2 = \frac{(u_x-v)^2 + (u^2-u_x^2)(1-v^2/c^2)}{(1-vu_x/c^2)^2}.$$

Thus:

$$1 - \frac{u'^2}{c^2} = 1 - \frac{\left(\dfrac{u_x}{c}-\dfrac{v}{c}\right)^2 + \left(\dfrac{u^2}{c^2}-\dfrac{u_x^2}{c^2}\right)\left(1-\dfrac{v^2}{c^2}\right)}{(1-vu_x/c^2)^2}$$

$$= \frac{1 - \dfrac{2vu_x}{c^2} + \dfrac{v^2u_x^2}{c^4} - \dfrac{u_x^2}{c^2} + \dfrac{2vu_x}{c^2} - \dfrac{v^2}{c^2} - \dfrac{u^2}{c^2} + \dfrac{u_x^2}{c^2} + \dfrac{v^2u^2}{c^4} - \dfrac{v^2u_x^2}{c^4}}{(1-vu_x/c^2)^2}$$

$$= \frac{1 - v^2/c^2 - u^2/c^2 + v^2u^2/c^4}{(1-vu_x/c^2)^2} = \frac{(1-v^2/c^2)(1-u^2/c^2)}{(1-vu_x/c^2)^2}.$$

Taking the square root:

$$\sqrt{(1-u'^2/c^2)} = \frac{\sqrt{[(1-v^2/c^2)(1-u^2/c^2)]}}{(1-vu_x/c^2)}. \tag{5.12}$$

100

Similarly, the inverse relation is:

$$\sqrt{(1-u^2/c^2)} = \frac{\sqrt{[(1-v^2/c^2)(1-u'^2/c^2)]}}{(1+vu_x'/c^2)}. \tag{5.13}$$

(b) *Transformation of momentum and energy*

In the laboratory system Σ, from equation (2.15) the momentum of a high speed particle is defined as:

$$p_x = mu_x = \frac{m_0 u_x}{\sqrt{(1-u^2/c^2)}}; \quad p_y = \frac{m_0 u_y}{\sqrt{(1-u^2/c^2)}}; \quad p_z = \frac{m_0 u_z}{\sqrt{(1-u^2/c^2)}}.$$

From equation (2.29) the total energy in Σ is:

$$E = mc^2 = \frac{m_0 c^2}{\sqrt{(1-u^2/c^2)}}.$$

The corresponding quantities in Σ' are defined as:

$$p_x' = m'u_x' = \frac{m_0 u_x'}{\sqrt{(1-u'^2/c^2)}}; \quad p_y' = \frac{m_0 u_y'}{\sqrt{(1-u'^2/c^2)}}; \quad p_z' = \frac{m_0 u_z'}{\sqrt{(1-u'^2/c^2)}}$$

and

$$E' = m'c^2 = \frac{m_0 c^2}{\sqrt{(1-u'^2/c^2)}}.$$

Substituting for $1/\sqrt{(1-u'^2/c^2)}$ from equation (5.12) and for u_x' from equation (5.6) into the expression for p_x', we obtain:

$$p_x' = \frac{m_0(1-vu_x/c^2)}{\sqrt{(1-u^2/c^2)}\sqrt{(1-v^2/c^2)}} \times \frac{(u_x-v)}{(1-vu_x/c^2)}$$

$$= \frac{m_0}{\sqrt{(1-u^2/c^2)}\sqrt{(1-v^2/c^2)}} (u_x-v) = \gamma(mu_x - mv),$$

where $\gamma = 1/\sqrt{(1-v^2/c^2)}$ and $m = m_0/\sqrt{(1-u^2/c^2)}$. But $mu_x = p_x$ and $m = E/c^2$. Hence:

$$p_x' = \gamma(p_x - vE/c^2).$$

Substituting for $1/\sqrt{(1-u'^2/c^2)}$ from equation (5.12) and for u_y' from equation (5.7) into the expression for p_y', one obtains:

$$p_y' = \frac{m_0 u_y'}{\sqrt{(1-u'^2/c^2)}} = \frac{m_0(1-vu_x/c^2)}{\sqrt{(1-u^2/c^2)}\sqrt{(1-v^2/c^2)}} \times \frac{u_y\sqrt{(1-v^2/c^2)}}{(1-vu_x/c^2)}$$

$$= \frac{m_0}{\sqrt{(1-u^2/c^2)}} u_y = p_y.$$

Similarly:

$$p_z' = p_z.$$

101

Using equation (5.12):

$$E' = m'c^2 = \frac{m_0 c^2}{\sqrt{(1 - u'^2/c^2)}} = \frac{m_0(1 - vu_x/c^2)c^2}{\sqrt{(1 - u^2/c^2)}\,\sqrt{(1 - v^2/c^2)}}$$

$$= \gamma m c^2 (1 - vu_x/c^2),$$

$$E' = \gamma(E - vp_x).$$

Collecting the transformations:

$$p_x' = \gamma(p_x - vE/c^2); \quad p_y' = p_y; \quad p_z' = p_z; \quad E' = \gamma(E - vp_x). \quad (5.14)$$

The inverse transformations are:

$$p_x = \gamma(p_x' + vE'/c^2); \quad p_y = p_y'; \quad p_z = p_z'; \quad E = \gamma(E' + vp_x'). \quad (5.15)$$

(c) *The force transformations*

It can be shown (Rosser[1]) that, if the force acting on a particle measured relative to Σ is **f** having components f_x, f_y and f_z, then the components of the force acting on the particle, measured relative to Σ', are:

$$f_x' = f_x - \frac{vu_y}{c^2(1 - vu_x/c^2)} f_y - \frac{vu_z}{c^2(1 - vu_x/c^2)} f_z, \quad (5.16)$$

$$f_y' = \frac{f_y}{\gamma(1 - vu_x/c^2)}, \quad (5.17)$$

$$f_z' = \frac{f_z}{\gamma(1 - vu_x/c^2)}. \quad (5.18)$$

5.5 *The centre of mass system**

(a) *Introduction*

To illustrate the use of the centre of mass system, the collision of two particles of equal rest masses m_0 will be considered. Let particle 1 have velocity u_1, momentum p_1, kinetic energy T_1 and total energy $E_1 = T_1 + m_0c^2$ relative to the laboratory system Σ before the collision, as shown in fig. 5.2 *a*. Let particle 1 move parallel to the x axis of Σ and collide with particle 2, which is stationary relative to the laboratory system Σ before the collision. Hence for particle 2, before the collision, we have relative to Σ, $u_2 = 0$, $p_2 = 0$, $T_2 = 0$ and $E_2 = m_0c^2$. The total linear momentum in the laboratory system Σ before the collision is $p_1 + p_2 = p_1 + 0 = p_1$. It is impossible for both the colliding particles to be at rest in the laboratory system Σ after the collision, since, even if the particles stick together, the particles must have a total momentum p_1 relative to the laboratory system Σ after the collision, if linear momentum is to be conserved relative to the laboratory system Σ. However, consider an inertial reference frame Σ' moving with uniform velocity v relative to Σ, such that the total linear momentum of the colliding particles relative to Σ' is zero before the collision. This is called the centre of mass system. If the total linear momentum in the centre of mass system is zero, particles 1 and 2 must approach each other with equal and opposite momenta and, since they have the same rest mass m_0, they must approach each other with equal and opposite speeds in the centre of mass system Σ', as shown in fig. 5.2 b. Since particle 2 is at rest in the laboratory system Σ before the collision, $u_x = 0$, so that according to equation

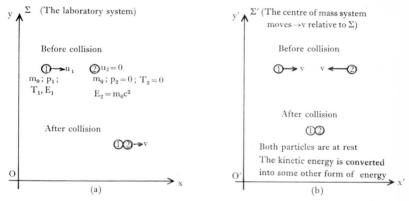

Figure 5.2. The inelastic collision of two particles having equal rest masses m_0. (a) Particle 2 is at rest in Σ before the collision, whilst particle 1 moves with velocity u_1, and momentum p_1 relative to Σ before the collision. (b) In the centre of mass system Σ' the total momentum is zero before the collision and particles 1 and 2 approach each other with equal speeds v relative to Σ'. It is possible for both particles to be at rest in the centre of mass system Σ' after the collision and still con-serve momentum. In this case particles 1 and 2 would have a speed v relative to the laboratory system Σ after the collision.

(5.6), before the collision the speed of particle 2 relative to the centre of mass system is:

$$u_x' = (u_x - v)/(1 - vu_x/c^2) = -v,$$

where v is the velocity of Σ' relative to Σ. If the magnitudes of the momenta of particles 1 and 2 in the centre of mass system Σ' are the same, then the speed of particle 1 in Σ' before the collision must be $+v$, as shown in fig. 5.2 b. The momenta of particles 1 and 2 before the collision are $m_0 v/\sqrt{(1-v^2/c^2)}$ and $-m_0 v/\sqrt{(1-v^2/c^2)}$ in the centre of mass system Σ', the total momentum being zero. In the centre of mass system Σ' it is possible for both particles 1 and 2 to come to rest in the collision and still conserve momentum, since the total momentum before the collision is zero in the centre of mass system. Thus it is possible to convert all the kinetic energy of the particles in the centre of mass system Σ' into some other form of energy, such as into heat or into creating new particles.

The speed v of the centre of mass system Σ' relative to the laboratory Σ will now be determined. Since before the collision the momentum and total energy of particle 1 relative to Σ are p_1 and E_1 respectively, according to equation (5.14) the momentum of particle 1 relative to the centre of mass system Σ' is:

$$p_1' = \gamma(p_1 - vE_1/c^2). \qquad (5.19)$$

Similarly, since $p_2 = 0$ and $E_2 = m_0c^2$ relative to Σ before the collision, relative to the centre of mass Σ', we have:

$$p_2' = \gamma(p_2 - vE_2/c^2) = -\gamma v m_0 c^2/c^2. \qquad (5.20)$$

Adding equations (5.19) and (5.20):

$$p_1' + p_2' = \gamma[p_1 - v(E_1 + m_0c^2)/c^2].$$

If the total momentum in the centre of mass system Σ' is to be zero before the collision, $p_1' + p_2'$ must be zero. Hence:

$$\gamma[p_1 - v(E_1 + m_0c^2)/c^2] = 0.$$

Hence the speed of the centre of mass system Σ' relative to the laboratory system Σ is:

$$v = \frac{c^2 p_1}{(E_1 + m_0c^2)} = \frac{c^2 p_1}{(T_1 + 2m_0c^2)}. \qquad (5.21)$$

Hence

$$1 - \frac{v^2}{c^2} = 1 - \frac{c^4 p_1{}^2}{c^2(E_1 + m_0c^2)^2}$$

$$= \frac{E_1{}^2 + 2E_1 m_0c^2 + m_0{}^2 c^4 - c^2 p_1{}^2}{(E_1 + m_0c^2)^2}.$$

From equation (2.30):

$$E_1{}^2 - c^2 p_1{}^2 = m_0{}^2 c^4.$$

Hence

$$1 - \frac{v^2}{c^2} = \frac{2E_1 m_0c^2 + 2m_0{}^2 c^4}{(E_1 + m_0c^2)^2} = \frac{2m_0c^2(E_1 + m_0c^2)}{(E_1 + m_0c^2)^2} = \frac{2m_0c^2}{(E_1 + m_0c^2)}$$

and

$$\frac{1}{\sqrt{(1 - v^2/c^2)}} = \frac{\sqrt{(E_1 + m_0c^2)}}{\sqrt{(2m_0c^2)}}. \qquad (5.22)$$

(b) Energy available in the centre of mass system

According to equation (2.29) the total energy of a particle of rest mass m_0 moving with speed v is $m_0c^2/\sqrt{(1 - v^2/c^2)}$, so that the total energy of each of the colliding particles in the centre of mass system Σ' before the collision is $m_0 c^2 / \sqrt{(1 - v^2/c^2)}$. Hence E', the total energy available in the centre of mass system Σ' before the collision, is given by:

$$E' = \frac{2m_0 c^2}{\sqrt{(1 - v^2/c^2)}} = 2m_0 c^2 \frac{\sqrt{(E_1 + m_0 c^2)}}{\sqrt{(2m_0 c^2)}},$$

where we have substituted for $1/\sqrt{(1 - v^2/c^2)}$ using equation (5.22). Hence:

$$E' = \sqrt{(2m_0 c^2 E_1 + 2m_0{}^2 c^4)}. \tag{5.23}$$

Since $E_1 = T_1 + m_0 c^2$:

$$E' = \sqrt{(2m_0 c^2 T_1 + 4m_0{}^2 c^4)}, \tag{5.24}$$

where T_1 and E_1 are the kinetic energy and the total energy of particle 1 in the *laboratory system* Σ before the collision. The *total* energy E' in the centre of mass system includes the total rest mass energy $2m_0 c^2$ of the colliding particles.

For simplicity we shall only consider cases in which the colliding particles are unchanged by the collision, but extra particles are produced. Firstly consider the case of a proton, denoted p^+, of kinetic energy T_1 in the laboratory system, which collides with a proton which is at rest in the laboratory system, producing a neutral π^0-meson according to the reaction:

$$\text{p}^+ + \text{p}^+ \rightarrow \text{p}^+ + \text{p}^+ + \pi^0.$$

The minimum value of T_1 to produce a π^0-meson of rest mass $m_\pi c^2 = 135 \cdot 0$ MeV will be calculated. The minimum value of T_1 is obtained when the two protons and the π^0-meson are at *rest* in the centre of mass system Σ' after the collision. In this case the only energy in the centre of mass system after the collision is equal to the total rest mass energies of the particles. Hence the total energy in the centre of mass after the collision is:

$$E' = 2m_0 c^2 + m_\pi c^2, \tag{5.25}$$

where $m_0 c^2 = 938$ MeV is the rest mass energy of a proton. Equating the right-hand sides of equations (5.24) and (5.25) and squaring, we have:

$$2m_0 c^2 T_1 + 4m_0{}^2 c^4 = 4m_0{}^2 c^4 + 4m_0 m_\pi c^4 + m_\pi{}^2 c^4,$$

$$T_1 = 2m_\pi c^2 + \frac{m_\pi{}^2 c^2}{2m_0} = m_\pi c^2 \left(2 + \frac{m_\pi}{2m_0}\right).$$

Substituting for $m_\pi c^2$ and $m_0 c^2$:

$$T_1 = 135 \left(2 + \frac{135}{2 \times 938}\right) = 280 \text{ MeV}.$$

In order to produce a π^0-meson in a proton–proton collision the proton accelerator must accelerate protons to kinetic energies above 280 MeV. The Berkeley frequency-modulated cyclotron, discussed in § 2.6, was able to do this and π-mesons were first produced in the laboratory using this machine.

The next generation of proton accelerators was designed to produce proton pairs according to the reaction:

$$p^+ + p^+ \to p^+ + p^+ + p^+ + p^-.$$

The negative proton, denoted p^-, is a particle of the same mass as an ordinary proton, but it has a negative charge of $-1\cdot602 \times 10^{-19}$ coulomb. The threshold kinetic energy for the above reaction will now be calculated. If all the four protons are at rest in the centre of mass system after the collision, then the total energy in the centre of mass system after the collision is:

$$E' = 4m_0c^2. \tag{5.26}$$

Equating the right-hand sides of equations (5.24) and (5.26) and squaring:

$$2T_1 m_0 c^2 + 4m_0{}^2 c^4 = 16m_0{}^2 c^4,$$

$$T_1 = 6m_0 c^2.$$

Since for a proton $m_0 c^2 = 938$ MeV, the threshold kinetic energy, that is, the minimum kinetic energy the incident proton must have in the laboratory system, is $5\cdot63$ GeV. The Berkeley Bevatron accelerated protons to kinetic energies of 6 GeV. The negative proton was discovered, using this machine by Chamberlain, Segrè, Wiegand and Ypsilantis in 1955.

The CERN proton synchrotron accelerates protons to energies up to nearly 30 GeV. At these energies $T_1(= 30 \text{ GeV}) \gg m_0 c^2$ $(= 0\cdot938 \text{ GeV})$. Hence equation (5.24) can be written in the approximate form:

$$E' \simeq \sqrt{(2m_0 c^2 T_1)}$$

for proton–proton collisions. At high energies the energy available in the centre of mass system goes up as the square root of the kinetic energy of the indicent particle in the laboratory system. For $T_1 = 30$ GeV and $m_0 c^2 = 0\cdot938$ GeV:

$$E' \sim \sqrt{(2 \times 0\cdot938 \times 30)} \sim 7\cdot5 \text{ GeV}.$$

If T_1 can be increased to 300 GeV using the machines now being designed:

$$E' \sim \sqrt{(2 \times 0\cdot938 \times 300)} \sim 23\cdot7 \text{ GeV}.$$

In this section we have given only a brief insight into high energy nuclear physics for the case of collisions between particles of equal

106

rest masses. The methods used, which are based on the equations developed in Chapter 2 and the transformations developed in § 5.4, can be applied in all cases, whatever the types of the colliding particles.

5.6 Electromagnetism via relativity*

Using the transformations of the theory of special relativity, it is possible to develop the laws of electromagnetism from Coulomb's law for the electric field due to a stationary point charge (reference: Rosser[2]). To give the reader an insight into this approach, we shall consider two simple examples. The following will be taken as *axiomatic* in this section.

(1) The force, velocity and co-ordinate transformations of the theory of special relativity.

(2) The principle of constant charge, according to which the numerical value of the total charge on a particle is the same in all inertial reference frames (cf. § 2.2).

(3) Coulomb's law for the electric field of a stationary point charge, such as an electron or a proton.

Consider two point charges q_1 and q_2. Let the charge q_1 move with *uniform* velocity v along the x axis of the laboratory system Σ as shown in both figs 5.3 a and 5.3 c. Let the charge q_1 be at O, the origin of Σ, at the time $t=0$. Two cases will be considered. In case (a), shown in fig. 5.3 a, the charge q_2 is at rest at the point P on the y axis of the laboratory system Σ. In case (b), shown in fig. 5.3 c, the charge q_2 is again at the point P on the y axis of the laboratory system Σ, but in this case q_2 is moving with uniform velocity u, parallel to the x axis of Σ at $t=0$. The force on the charge q_2 due to the moving charge q_1, measured in the laboratory system Σ at the time $t=0$ when q_1 is at the origin, will be calculated in both cases.

Case (a)

In the inertial reference frame Σ', which is moving with uniform velocity v relative to the laboratory system Σ along their common x axis, the charge q_1 is at rest at the origin O$'$, as shown in fig. 5.3 b. Corresponding to the position of q_2 at P in fig. 5.3 a at $x=0$, $y=y$ at $t=0$ in Σ, according to the Lorentz transformations, the charge q_2 is at P$'$ in fig. 5.3 b, which has coordinates $x'=0$, $y'=y$ at $t'=0$ relative to Σ'. According to Coulomb's law, relative to Σ', the electric field at P$'$ due to the charge q_1, which is at rest at the origin of Σ' is, since $y'=y$:

$$E_1' = \frac{q_1}{4\pi\epsilon_0 y'^2} = \frac{q_1}{4\pi\epsilon_0 y^2}.$$

Since the charge q_1 is at rest relative to Σ', it does not give rise to a magnetic field relative to Σ'. Hence, in Σ', the only force acting on the charge q_2 is an electric force given by equation (2.1). Hence,

107

relative to Σ', the total force acting on q_2 is in the y' direction and is equal to:

$$f_y' = q_2 E_1' = \frac{q_1 q_2}{4\pi \epsilon_0 y^2}. \qquad (5.27)$$

Rearranging equation (5.17), we have:

$$f_y = \gamma(1 - vu_x/c^2) f_y'. \qquad (5.28)$$

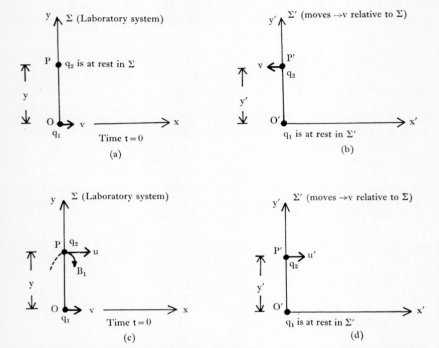

Figure 5.3. (a) The charge q_1 moves with uniform velocity v along the x axis of Σ, whereas q_2 is at rest in Σ. (b) The charge q_1 is at rest in Σ'. (c) The charge q_1 moves with uniform velocity v along the x axis of Σ, whereas q_2 moves with uniform velocity u parallel to the x axis of Σ. (d) The charge q_1 is at rest in Σ'.

Substituting for f_y' from equation (5.27), remembering that, for the case shown in fig. 5.3 a, the velocity of q_2 relative to Σ is zero, so that $u_x = 0$:

$$f_y = \gamma f_y' = \frac{\gamma q_1 q_2}{4\pi \epsilon_0 y^2} = q_2 \left(\frac{q_1}{4\pi \epsilon_0 y^2 (1 - v^2/c^2)^{1/2}} \right). \qquad (5.29)$$

The charge q_2 can be treated as a test charge to measure the electric and magnetic fields due to the moving charge q_1 in the laboratory

108

system Σ. Since q_2 is at rest in Σ, it follows from equation (2.2) that there is no magnetic force on q_2 in Σ. Hence equation (5.29) can be rewritten in the form:

$$f_y = q_2 E_1,$$

where

$$E_1 = q_1\gamma/4\pi\epsilon_0 y^2 \qquad (5.30)$$

is the electric field in Σ, due to the charge q_1, which is moving with uniform velocity v in the laboratory system Σ.

Case (b)

The charge q_1 is again at rest at O' the origin of the inertial reference frame Σ', which moves with uniform velocity v relative to the laboratory system Σ, as shown in fig. 5.3 *d*. Hence, relative to Σ', the charge q_1 gives rise to no magnetic field and, relative to Σ', the only force acting on q_2 is the electric force given by equation (5.27). In case (b) the velocity of q_2 relative to Σ is not zero, but q_2 has a velocity u in the x direction, so that, relative to Σ, $u_x = u$. Hence in this case, using equation (5.27), equation (5.28) becomes:

$$\left.\begin{array}{l} f_y = \gamma(1 - vu/c^2)q_1 q_2/4\pi\epsilon_0 y^2, \\[2mm] f_y = q_2\left(\dfrac{q_1\gamma}{4\pi\epsilon_0 y^2}\right) - q_2 u\left(\dfrac{q_1 v\gamma}{4\pi\epsilon_0 c^2 y^2}\right) \end{array}\right\} \qquad (5.31)$$

The first term on the right-hand side of equation (5.31) is independent of u, the velocity of the test charge q_2. It is the electric force acting on q_2 and is equal to $q_2 E_1$, where E_1 is the electric field due to q_1 in the laboratory system Σ, and is again given by equation (5.30). The second term on the right-hand side of equation (5.31) is an extra term in the negative y direction, compared with equation (5.29). It depends on u, the velocity of the test charge q_2 relative to the laboratory system Σ. Relative to the laboratory system Σ, this extra force can be interpreted as a magnetic force:

$$(f_{\text{mag}})_y = -q_2 u\left(\dfrac{q_1 v\gamma}{4\pi\epsilon_0 c^2 y^2}\right) = -q_2 u B_1, \qquad (5.32)$$

where

$$B_1 = q_1 v\gamma/4\pi\epsilon_0 c^2 y^2 \qquad (5.33)$$

is the magnetic field at P in fig. 5.3 *c*, in the laboratory system Σ, due to the charge q_1 which is moving with uniform velocity v. At P the direction of this magnetic field is in the $+z$ direction upwards towards the reader as shown in fig. 5.3 *c*. This direction is consistent with the normal right-handed corkscrew rule. The magnetic force on

109

q_2 should be consistent with equation (2.2). In this case the angle between u and B is 90°, so that in equation (2.2), sin $\alpha = 1$. Equations (2.2) and (5.33) give equation (5.32). The reader can check that the direction of the magnetic force on q_2 in fig. 5.3 c is consistent with the left-hand motor rule, illustrated in fig. 2.1 b. Thus it is possible to calculate the magnetic field of a moving charge and the magnetic forces between moving charges from Coulomb's law using relativity theory. Notice in equation (5.31), the ratio of the magnetic force on q_2 to the electric force on q_2 is uv/c^2. Since both u and v must be less than c, the magnitude of the magnetic force on q_2 due to q_1 is always less than the electric force on q_2.

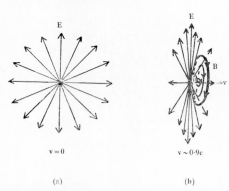

Figure 5.4. (*a*) The electric field of a stationary positive charge is spherically symmetric. (*b*) If the charge is moving with uniform velocity v, the electric intensity still diverges from the position of the charge, but the electric intensity is increased in the direction perpendicular to v, and decreased in the directions parallel and antiparallel to v. The magnetic field lines in one plane are sketched. The magnetic field lines are circles concentric with the direction of v.

Case (c) *General case of a charge moving with uniform velocity*

Using methods similar to the above, it can be shown (Rosser[2]) that the electric field E and the magnetic field B at a distance r from a point charge of magnitude q moving with uniform velocity v are given by:

$$E = \frac{q(1 - v^2/c^2)}{4\pi \epsilon_0 r^2 \{1 - (v^2/c^2) \sin^2 \theta\}^{3/2}}, \tag{5.34}$$

$$B = \frac{qv(1 - v^2/c^2) \sin \theta}{4\pi \epsilon_0 c^2 r^2 \{1 - (v^2/c^2) \sin^2 \theta\}^{3/2}}, \tag{5.35}$$

where θ is the angle between v and r. When $v = 0$ equation (5.34) reduces to:

$$E = q/4\pi \epsilon_0 r^2,$$

which is Coulomb's law. When an electric charge q is moving with *uniform* velocity, its electric field lines diverge radially from the position of the charge, just as in the electrostatic case. The electric intensity in different directions is, however, different in the two cases, as shown in figs $5.4\,a$ and $5.4\,b$. When $\theta=0$ in fig. $5.4\,b$, from equation (5.34), $E=q(1-v^2/c^2)/4\pi\epsilon_0 r^2$, so that when the charge q is moving, the electric intensity in the direction of motion is reduced compared with the electrostatic value of $q/4\pi\epsilon_0 r^2$. When $\theta=90°$, sin $\theta=1$ and $E=q/4\pi\epsilon_0 r^2(1-v^2/c^2)^{1/2}$, in agreement with equation (5.30). Hence the electric intensity is increased in the direction perpendicular to v compared with the electrostatic case, as shown in fig. $5.4\,b$.

The magnetic field lines of the moving charge are circles in planes perpendicular to v, the centres of the circles being along the direction of motion of the charge q. The magnetic field lines in one plane are sketched in fig. $5.4\,b$. The direction of the magnetic field is consistent with the right-handed corkscrew rule. If $\theta=90°$, equation (5.35) reduces to equation (5.33). If $v\ll c$ equation (5.35) becomes:

$$B\sim\frac{qv\sin\theta}{4\pi\epsilon_0 c^2 r^2},\qquad(5.36)$$

which is the expression for the Biot–Savart law. When a steady conduction current flows in a wire, the current is due to the motions of the conduction electrons. The velocities of the conduction electrons are much less than c, and equation (5.36), the Biot–Savart law, can be used to calculate the magnetic field due to a steady conduction current.

Only a brief insight into relativity and electromagnetism has been given for the case of charges moving with uniform velocities, so as to illustrate how electrostatics and electromagnetism are related. For further details, see Rosser[2].

References
1. ROSSER, W. G. V., *Introductory Relativity*, (Butterworths, London, 1967), p. 145.
2. ROSSER, W. G. V., *Classical Electromagnetism Via Relativity* (Butterworths, London, 1968).

Problems
(Assume that the velocity of light $c=3\cdot00\times10^8$ m s⁻¹)

5.1. A train is passing through a station at a speed of 20 m s⁻¹ A marble is rolled along the floor of one of the compartments with a velocity of 10 m s⁻¹ relative to the train. Calculate the speed of the marble relative to an observer standing on the platform
   (a) if the marble rolls in the direction of motion of the train,
   (b) if the marble rolls perpendicular to the direction of motion of the train as measured by a passenger on the train.

5.2. A radioactive nucleus is moving with a velocity $c/10$ relative to the laboratory, when it emits a β-particle with a velocity $0 \cdot 9c$ relative to the co-ordinate system in which the decaying nucleus is at rest. What is the velocity and direction of the β-particle relative to the laboratory if, relative to the radioactive nucleus, it is emitted
(a) in the direction of motion of the nucleus relative to the laboratory,
(b) perpendicular to the direction of motion?

5.3. A radioactive nucleus which is at rest in Σ', but is moving with velocity $c/6$ along the x axis of Σ, emits a β-particle of velocity $0 \cdot 8c$ (relative to Σ') at an angle of $45°$ to the x' axis of Σ'. What is the velocity of the β-particle relative to an observer at rest in an inertial reference frame Σ'' going at a velocity of $c/2$ along the negative x axis of Σ? (Hint: Find the velocity of Σ' relative to Σ'', and then apply the velocity transformations between Σ' and Σ''.)

5.4. An observer at rest in Σ' moves along the x axis of Σ with velocity v. He measures the length of a body of proper length l_0, which is moving with velocity u along the x axis of Σ. Show that he measures the length to be:

$$l_0 \sqrt{\left(\frac{(c^2 - v^2)(c^2 - u^2)}{(c^2 - uv)^2} \right)}.$$

[Hint: He measures the length to be $l_0 \sqrt{(1 - u'^2/c^2)}$. Use equation (5.12.)]

accelerating reference frames: the principle of equivalence

Throughout this chapter, the discussion will be confined to *low* enough speeds (very much less than the speed of light) such that Newtonian mechanics and the Galilean transformations would be satisfactory approximations at such speeds in inertial reference frames.

6.1 *Linearly accelerating reference frames*

It was discussed in Chapter 1 how, if the ship illustrated in fig. 1.1 is moving with uniform velocity relative to the earth, it is possible to play table tennis in the games room.   One does not have to stop and think, before each stroke, which way the ship is moving.   One just plays one's normal game.   In fact, if the ship were moving with uniform velocity relative to the earth, one could not say that the ship was moving without looking at something external to the ship.   However, if the ship encounters stormy seas one would soon realize that it would be impossible to play a normal game of table tennis, and that one would have to try to allow for the pitching and rolling of the ship. It is possible to say when the ship is accelerating relative to the earth.

Let the ship illustrated previously in fig. 1.1 be accelerating relative to the earth with acceleration **a**, when the mass m is dropped, as shown in fig. 6.1.   Relative to the accelerating ship the mass m appears to have an acceleration $-$ **a** and the mass m appears to be deflected away from the ' vertical ', as shown in fig. 6.1.

If a *frictionless* mass were given a push on the perfectly *smooth* deck of a ship moving with *uniform* velocity, the mass would move in a straight line relative to the ship and relative to the earth.   If the ship were *accelerating* with acceleration **a** relative to the earth, if there were *no* friction at all between the moving mass and the deck of the ship, the mass would still move in a straight line relative to the earth.   However, unless the mass were moving in a direction parallel to the acceleration of the ship, the mass would *not* move in a straight line relative to the ship, but would have an acceleration $-$ **a** relative to the accelerating ship.   Hence, whatever its mass, a body, not acted upon by any forces in the horizontal plane, has an acceleration $-$ **a** relative to a ship which has an acceleration $+$ **a** relative to the earth.   Newton's first law would *not* be valid relative to the accelerating ship, and the accelerating ship would *not* be an inertial reference frame, and Newton's laws of motion should *not* be applied relative to the accelerating ship.   In practice one can get over this by saying that in accelerating non-inertial reference frames *all* bodies are acted upon by *fictitious* inertial 'forces'. For example, it would be assumed that a fictitious 'force' $-m\mathbf{a}$ acted on the falling mass m in fig. 6.1. Relative to the earth this fictitious 'force' need not be introduced. It is a convenient fiddle to introduce the *fictitious* inertial 'forces' in the accelerating reference frame, so that Newton's laws can be applied in conditions where strictly Newton's laws should not really be applied. The fictitious inertial 'forces' in accelerating reference frames are always opposite in direction to the direction of the acceleration of the accelerating reference frame relative to an inertial reference frame. In fig. 6.1, in the accelerating reference frame in which the ship is at rest, one would say that the falling mass m is deflected from the 'vertical' by the inertial 'force' $-m\mathbf{a}$. This fictitious 'force' gives no acceleration relative to the earth.

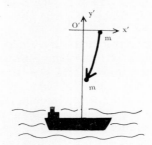

Figure 6.1. Relative to the accelerating ship, the mass m, which is dropped from the mast, appears to be deflected away from the vertical in a direction opposite to the direction in which the ship is accelerating relative to the earth. This deflection is interpreted in terms of a fictitious inertial force $-m\mathbf{a}$ in the accelerating reference frame.

The reader is probably more familiar with fictitious inertial 'forces' in rotating reference frames. Due to the earth's rotation, the laboratory system is not an inertial reference frame, and strictly one should not apply Newton's laws in the laboratory system. It would be a great inconvenience always to have to apply Newton's laws in a system at rest relative to the fixed stars. It will be shown that one can still apply Newton's laws within a frame of reference fixed on the rotating earth, if one does the fiddle of introducing the fictitious centrifugal 'force' and the fictitious Coriolis 'force'.

6.2 *Rotating reference frames*
Generally phenomena will be interpreted initially from the view-

114

point of an inertial reference frame in which Newton's laws can be applied. It will then be seen how the results can be interpreted by an observer at rest in a reference frame rotating relative to that inertial reference frame. We shall start with a simple example to show that Newton's first law is *not* valid in a rotating reference frame.

Consider a roundabout which is rotating with uniform angular velocity ω relative to the laboratory system Σ, as shown in fig. 6.2a. Let a bullet be fired horizontally to pass over the axis of rotation of the roundabout, from a gun which is at *rest* relative to the laboratory system Σ. Relative to the laboratory system Σ, since it is not acted upon by any forces in the horizontal plane, the bullet moves in a straight line

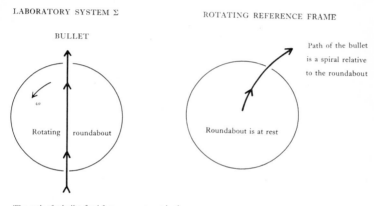

LABORATORY SYSTEM Σ

BULLET

Rotating roundabout

The path of a bullet fired from a gun at rest in the laboratory system is a straight line in the laboratory system

(a)

ROTATING REFERENCE FRAME

Path of the bullet is a spiral relative to the roundabout

Roundabout is at rest

(b)

Figure 6.2. (*a*) The roundabout is rotating in the laboratory system Σ. A bullet fired from a gun at rest in the laboratory system Σ travels in a straight line relative to Σ. (*b*) Relative to the rotating reference frame, in which the rotating roundabout is at rest, the path of the bullet is not a straight line, even though no forces are applied in the horizontal direction. Newton's first law is not valid in the rotating reference frame, since a particle not acted upon by any forces does not travel in a straight line in the rotating reference frame.

relative to the horizontal plane, as shown in fig. 6.2 *a*. (The vertical motion under gravity is being neglected.) However, relative to the rotating reference frame in which the roundabout is at rest, the bullet appears to go in a curved path, as shown in fig. 6.2 *b*. When, after the bullet has passed over the axis of rotation, the roundabout has made one further revolution relative to the laboratory system Σ, the bullet will then appear to have gone in a spiral path through 360° in the opposite direction relative to the rotating roundabout, since

115

it passed the axis of rotation. Thus Newton's first law is *not* valid in the rotating reference frame in which the roundabout is at rest, which is therefore *not* an inertial reference frame. Strictly, Newton's laws should not be applied in the rotating reference frame. It will now be illustrated how Newton's laws can however be used in a rotating reference frame, if we invent the *fictitious* centrifugal ' force ' and the *fictitious* Coriolis ' force '.

Consider a particle P, of mass m which is attached by a spring balance to a point O. Let P be at a distance r from O. Let the particle P rotate with uniform angular velocity ω about an axis through O, as shown in fig. 6.3 *a*. Relative to the laboratory system Σ, which at present will be considered as an approximate inertial reference frame, since the particle P is moving in a circle, it has an acceleration $r\omega^2$ directed towards O. In order to give rise to this acceleration, there must be a force equal to $mr\omega^2$ acting on the particle P. This centripetal force is due to the pull of the spring balance on the mass m. It can be measured by the spring balance. This centripetal force is necessary, since the particle P is tending to carry on in a straight line relative to the laboratory system, Σ, that is, it is tending to fly off at a tangent to the circular motion. The force exerted on P by the spring balance keeps the particle P moving in a circle.

Now consider a co-ordinate system, denoted by A, which is rotating relative to the laboratory system, such that, relative to A, the particle P is at rest, as shown in fig. 6.3 *b*. There is still a centripetal force equal to $mr\omega^2$ acting on P and directed towards O, due to the tension in the spring balance. An observer at rest in A could read the tension in the spring balance. If this were the only force acting on P, if Newton's laws were applicable in the rotating reference frame A, this force should lead to an acceleration of the particle P towards O. In fact the particle P is at rest relative to the rotating reference frame A. One overcomes this difficulty by inventing a *fictitious* ' force ', and saying that a ' force ' $mr\omega^2$ acts *outwards* on the particle P. Relative to the rotating reference frame A, there would be no resultant ' force ' on P, and P should remain at rest in the rotating reference frame. This fictitious outward ' force ' is generally known as the *centrifugal* '*force*'. It produces no acceleration relative to the fixed stars. It is only introduced in rotating reference frames.

Consider a particle of mass m at rest on the surface of the earth at a point having latitude λ in the Northern Hemisphere, as shown in fig. 6.3 *c*. Relative to the fixed stars the particle has a tendency to fly off in a straight line. Relative to the rotating earth, if one wants to apply Newton's laws, one has to introduce the fictitious centrifugal ' force '. If $R = r \cos \lambda$ is the perpendicular distance of the mass m from the axis of rotation (r is the radius of the earth) then the fictitious ' centrifugal force ' acting on the mass m would be $mR\omega^2 = mr\omega^2 \cos \lambda$. This force acts outwards, as shown in fig. 6.3 *c*. The other force

116

acting on the mass m is the gravitational force of attraction due to the earth. This acts downwards towards the centre of the earth. If M is the mass of the earth, G the gravitational constant and g_0 the value the acceleration due to gravity would have at the surface of the earth, if the earth did not rotate, then from Newton's law of universal gravitation we have: $mg_0 = G(mM/r^2)$.

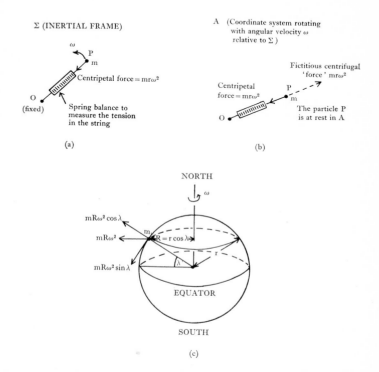

Figure 6.3. (a) The particle P rotates with uniform angular velocity ω on a smooth horizontal table. (b) In the co-ordinate system A, which rotates with P, a fictitious centrifugal 'force' $mr\omega^2$ must be introduced, if one wants to apply Newton's laws of motion in the rotating co-ordinate system. (c) The centrifugal 'force' $mR\omega^2$ relative to the earth reduces the effective value of the acceleration due to gravity.

If we choose as our standard of rest a reference frame rotating with the earth, we must introduce the *fictitious* centrifugal 'force' $mR\omega^2 = mr\omega^2 \cos \lambda$ acting outwards from the axis of rotation of the earth, as shown in fig. 6.3 c. The resultant 'force' on the mass m would have components $(mg_0 - mr\omega^2 \cos^2 \lambda)$ in the direction radially

117

inwards, towards the centre of the earth, and a tangential component $mr\omega^2 \cos \lambda \sin \lambda$ acting in the south direction in fig. 6.3 c. A plumb line would set in the direction of the resultant of these forces and would not point towards the centre of the earth. The terms associated with the centrifugal 'force' are small compared with the gravitational attraction of the earth, so that in practice the deviation of a plumb line from the radial direction towards the centre of the earth is very small. The effective value of the acceleration due to gravity in the direction towards the centre of the earth and denoted by g is given by:

$$mg = mg_0 - mr\omega^2 \cos^2 \lambda. \qquad (6.1)$$

At the North Pole, $\lambda = 90°$, $\cos \lambda = 0$ and $g = g_0$. At the Equator $\lambda = 0$, $\cos \lambda = 1$ and the effective value of g is $(g_0 - r\omega^2)$. Therefore the fractional change in g between the Pole and the Equator is $\omega^2 r/g_0$. Now, since the earth turns through an angle of 2π radian about its axis every 24 hours, $\omega = 7 \cdot 29 \times 10^{-5}$ radian per second. The radius of the earth is $6 \cdot 371 \times 10^6$ metre. Taking $g_0 = 9 \cdot 81$ metre per second^2, we find the calculated fractional variation of g to be $0 \cdot 3\%$. This is in agreement with experiment. Relative to the fixed stars, the effect is due to the tendency of a particle rotating with the earth to fly off at a tangent and travel in a straight line relative to the fixed stars. This effect is bigger the bigger the distance of the particle from the axis of rotation of the earth. This leads to the lower *measured* value of g at low latitudes and the bulging of the earth near the Equator.

There is another important effect which arises when a body *moves* relative to a rotating reference frame, such as a rotating roundabout. Consider a roundabout rotating with uniform angular velocity ω relative to the laboratory system Σ, as shown in fig. 6.4 a. Let somebody at rest on the roundabout at a distance r from O, the axis of rotation, shoot a gun, *fixed to the roundabout*, such that a bullet is fired radially outwards with velocity u' along the line OP. Relative to the laboratory system Σ, in addition to the velocity u' in the radial direction, the bullet has a velocity $r\omega$ in a direction perpendicular to the radius OP of the roundabout, as shown in fig. 6.4 a. Relative to the laboratory system Σ, the bullet goes in a *straight line* at an angle θ, given by $\tan \theta = (r\omega/u')$, to the direction of aim, as shown in fig. 6.4 a. Let the bullet move a distance Δr outwards to $(r + \Delta r)$ in a time Δt. An observer, at rest on the roundabout at a point Q at a distance $(r + \Delta r)$ from the axis of rotation on the line OP moves with a linear speed $(r + \Delta r)\omega$ relative to the laboratory system Σ, and, due to this greater speed perpendicular to OP, the observer at Q at $(r + \Delta r)$ from O will have moved farther to the left in the time Δt than the bullet, relative to the laboratory system Σ. The line OP joining the axis of rotation to the observer at Q rotates to the left to OP' in the time

118

Δt, such that the observer is then at Q' in fig. $6.4\,a$. Hence the bullet fired outwards in fig. $6.4\,a$ will appear to be deflected to the right relative to the observer at rest on the line OP on the rotating roundabout, as shown in fig. $6.4\,b$. The centrifugal 'force' $mr\omega^2$ acts radially outwards in this example and does not contribute to the initial deflection of the bullet in a direction perpendicular to OP, relative to the rotating roundabout. Another fictitious force, the *Coriolis* 'force' is invented to interpret the deflection of the bullet in a direction perpendicular to OP in the rotating reference frame.

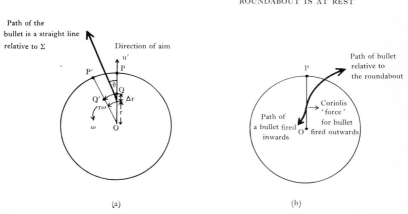

LABORATORY SYSTEM Σ

ROTATING REFERENCE FRAME

ROUNDABOUT IS AT REST

(a)

(b)

Figure 6.4. (*a*) Laboratory system Σ. A bullet is fired outwards by a gun fixed to the rotating roundabout. (*b*) Relative to the rotating roundabout the bullet appears to be deflected from a straight line. In the rotating reference frame this deflection is interpreted in terms of the Coriolis and centrifugal 'forces'. The cases of bullets fired outwards and inwards towards the axis of rotation are shown in (*b*).

It can be shown that, if the bullet is fired inwards towards the axis of rotation, the deflection due to the Coriolis 'force' is as shown in fig. $6.4\,b$. Notice, when the roundabout is rotating in the direction shown in fig. $6.4\,a$, if we look along the direction of motion of the bullet, the bullet is deflected to the right due to the Coriolis 'force', whether it is fired outwards or inwards towards the axis of rotation. Newton's laws of motion are not really valid on a rotating roundabout. They can be used, if the *fictitious* centrifugal and Coriolis 'forces' are introduced to interpret the 'deflections' of the bullet relative to the rotating roundabout.

When one is walking radially outwards (or inwards) on a rotating roundabout one feels jerks on one's feet as one changes one's distance from the axis of rotation. The force on one's front foot as one walks

119

outwards on the roundabout is due to the fact that the linear speed $r\omega$ of a point on the surface of the roundabout increases with increasing distance r from the axis of rotation. There is also a contribution to these forces due to the fact that the direction of this linear speed is changing with time, due to the rotation of the roundabout. These forces are real forces due to the friction between the shoes and the roundabout. (One would soon topple all over the place on a smooth roundabout.) These real forces compensate the Coriolis ' force ' and prevent deflections, relative to the rotating roundabout, of the type shown in fig. 6.4 b. The centrifugal ' force ' is ' felt ' as a tendency to move radially outwards on a rotating roundabout, and is present whether one is moving or one is stationary on the roundabout.

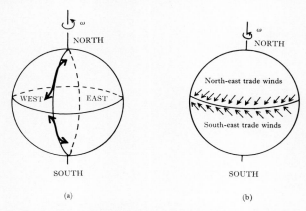

Figure 6.5. (*a*) Due to the Coriolis ' force ', ballistic missiles are deflected relative to the earth as shown. (*b*) The heating of air near the Equator leads to winds towards the Equator from more temperate regions. These winds are deflected by the Coriolis ' force '.

A few practical examples of the effect of Coriolis ' forces ' will now be considered. Strictly, the laboratory system is not an inertial reference frame, since the earth is rotating about its axis with angular velocity $\omega = 7 \cdot 29 \times 10^{-5}$ radian per second. The direction of rotation of the roundabout in fig. 6.4 a is the same as the direction of rotation of the earth, as viewed from above the North Pole. (Compare figs 6.4 a and 6.5 a. The reader can also check this by looking at a globe.) If a naval ship in the Northern Hemisphere fires a missile in the north direction, due to the rotation of the earth, the missile is deflected to the right, relative to the earth, as viewed along the direction of motion of the missile. In this case the missile is deflected to the east, as illustrated in fig. 6.5 a. Relative to a reference frame at rest relative to the earth, this deflection is attributed to the Coriolis ' force ' acting on the missile. If the missile is fired in the south direction, from a

120

point in the Northern Hemisphere, since relative to the rotating reference frame the missile is deflected to the right (when looking along the path of the missile), the missile is deflected to the west relative to the earth, as shown in fig. 6.5 *a*. These effects are important. For example, if the missile were fired at an elevation of 45° with an initial velocity of 550 metre per second in the north direction from a point having latitude $\lambda = 45°$ north, the range of the missile would be about 30 kilometre, and it would be deflected to the east by about 200 metre. This would be more than enough to miss the target. Thus naval gunners must allow for the centrifugal and Coriolis ' forces '. It is left as an exercise for the reader to show that, in the Southern Hemisphere, the missiles would be deflected, as shown in fig. 6.5 *a*. (A naval gunner must be careful to remember which side of the Equator he is.)

Coriolis ' forces ' give rise to the trade winds in equatorial regions. The air is generally warmer near the Equator than at higher latitudes. The air near the Equator rises vertically and is replaced by winds blowing towards the Equator from more temperate regions. Due to the Coriolis ' force ' that must be introduced in the rotating reference frame in which the earth is at rest, by analogy with fig. 6.5 *a*, the reader can see that these winds will come from the north-east in the Northern Hemisphere and from the south-east in the Southern Hemisphere, as shown in fig. 6.5 *b*.

The reader has probably seen weather charts on television, showing high and low pressure systems. If we have a low pressure system, the winds blow inwards towards the centre of the low pressure system to try and equalize the air pressures. However, relative to the reference frame rotating with the earth, these winds are deflected to the right in the Northern Hemisphere by the Coriolis ' force ', as shown in fig. 6.6 *a* for the case of the Northern Hemisphere. Thus, due to the Coriolis ' force ', the air tends to circulate in an anti-clockwise direction around low pressure systems in the Northern Hemisphere, as shown in fig. 6.6 *a*. The winds go predominantly around, not across, low pressure systems, which can then persist for much longer times than they would have done if there were no Coriolis ' force ', that is, if the earth did not rotate. With a high pressure system (or anticyclone) the winds tend to blow outwards from the centre of the anticyclone. Due to the Coriolis ' force ' the winds are deflected to the right in the Northern Hemisphere, as shown in fig. 6.6 *b*, such that, in the Northern Hemisphere, the winds go in a clockwise direction around anticyclones. It is left as an exercise for the reader to show that in the Southern Hemisphere the winds go in a clockwise direction around low pressure systems, and in an anticlockwise direction around high pressure systems.

The centrifugal and Coriolis ' forces ' are fictitious in the sense that they produce no accelerations relative to an inertial reference frame. In fact *they must not be introduced in inertial reference frames*. They

121

are inventions to enable us to use Newton's laws of motion in rotating reference frames, in conditions where Newton's laws of motion are not really applicable. However, we live in a rotating reference frame, though we tend to forget it. The deflection of ballistic missiles, the directions of the winds, etc. appear real enough to us in our rotating reference frame, simply because we are accustomed to thinking in terms of Newton's laws, ignoring the rotation of the earth. Even though they are fictional, it is easier to tell a naval gunner to introduce the Coriolis and centrifugal ' forces ' and to use Newton's laws of motion rather than be pedantic and say that Newton's laws should not be applied, since they are not valid relative to the rotating earth, and that the gunner should relate everything to an inertial reference frame at rest relative to the fixed stars.

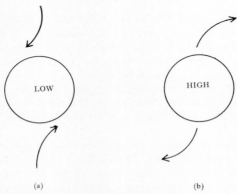

(a)   (b)

Figure 6.6.   (a) Due to Coriolis ' forces ', the winds towards the centre of a low pressure system are deflected such that they go around the low pressure system in an anti-clockwise direction in the Northern Hemisphere. (b) The winds go in a clockwise direction around a high pressure system in the Northern Hemisphere.

After developing his theory of special relativity Einstein went on to discuss the question whether the laws of physics can be expressed in the same form in all reference frames, whether they are accelerating or not. This led to the appreciation of the close relationship between gravity and accelerating reference frames. This will now be illustrated in terms of the principle of equivalence.

6.3. The principle of equivalence

(a) Introduction

Let an astronaut be in a spaceship which is at rest in an inertial reference frame Σ in a region of outer space where there is *no* gravitational field, as shown in fig. 6.7 a. There are no windows in the spaceship, so that the astronaut cannot look out, and it is soundproof

122

so that he cannot hear anything.   Everything inside the spaceship is
in a state of weightlessness.   Now let a signal from the earth fire some
small rockets, so that the spaceship starts moving with uniform acceleration **a** relative to Σ.   Consider what happens from the viewpoint of the

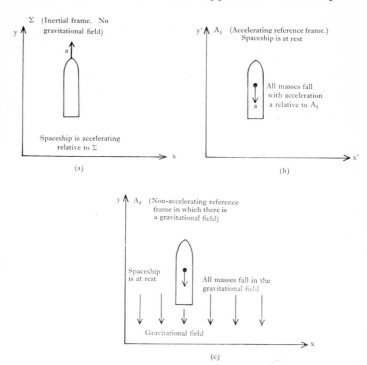

Figure 6.7.   (a) The spaceship has uniform acceleration *a* relative to the inertial
reference frame Σ.   (b) The accelerating reference frame A_1 is the reference
frame in which the spaceship is at rest. A_1 has an acceleration *a*
relative to Σ.   All bodies not acted upon by any forces accelerate
downwards in A_1.   According to the principle of equivalence, measurements carried out in A_1 yield the same results as experiments carried
out in the non-accelerating reference frame A_2 shown in (c), in which
there is a gravitational field, opposite in direction to *a* and of such
a strength that the acceleration due to the force of gravity is numerically
equal to *a*.

reference frame A_1, shown in fig. 6.7 *b*, in which the accelerating
spaceship is at rest.   The astronaut is no longer in a state of weightlessness.   If he holds up an apple and lets it go, the apple will continue
to move with the same velocity, relative to Σ, as the spaceship had
when the apple was released.   Hence, relative to the spaceship, the
apple falls with uniform acceleration.   If two apples of different

123

masses were dropped at the same time, they would have the same acceleration $-\mathbf{a}$ relative to the accelerating spaceship, so that all falling bodies would have the same acceleration relative to the spaceship. If a spring balance, with a mass attached to it, is attached to the roof of the spaceship, the mass would remain at rest relative to Σ when the spaceship started accelerating, unless the spring in the spring balance were under sufficient tension to give the mass the same acceleration as the spaceship has relative to Σ. If the acceleration of the spaceship were numerically equal to g, the acceleration due to gravity at the surface of the earth, the conditions in the spaceship would be similar to the conditions in a spaceship at rest on the surface of the earth. How should the astronaut interpret the results in the accelerating spaceship, if he did not know that rockets were accelerating the spaceship? He would be just as justified as Newton was in concluding from the observations on a falling apple, that he was in a gravitational field. The astronaut could interpret the tension in the spring balance in terms of a force of gravitational origin on the mass, when it is in a gravitational field. In fact, the astronaut inside the spaceship could not tell the difference between times when he was accelerating relative to the fixed stars, as shown in fig. 6.7 b, and times when the spaceship was at rest relative to the fixed stars, and a massive body was underneath the spaceship, giving rise to a gravitational field, as shown in fig. 6.7 c. In 1911 Einstein proposed that the co-ordinate system A_1, shown in fig. 6.7 b, which is accelerating with uniform acceleration $\mathbf{a}$ relative to the fixed stars and the co-ordinate system A_2 shown in fig. 6.7 c, which is at rest in a gravitational field in which the acceleration due to gravity is numerically equal to a, are exactly equivalent. That is, experiments carried out under identical conditions in A_1 and A_2 should give the same numerical results. This is the *principle of equivalence*.

(b) *The rates of clocks in gravitational fields and the gravitational red shift*

The principle of equivalence will now be used to calculate the rates of clocks in gravitational fields. It should be remembered that, throughout this chapter, it is being assumed that all speeds are very much less than the speed of light, and the formulae of Newtonian mechanics will be used.

Consider a spaceship of length h moving with uniform acceleration $\mathbf{a}$ relative to an inertial reference frame Σ in which there is no gravitational field, as shown in fig. 6.8 a. Let the spaceship be instantaneously at rest in Σ at the time $t = 0$. Let two clocks of identical construction, labelled 1 and 2, be fixed to the spaceship, with Clock 1 at the rear and Clock 2 at the front end of the spaceship. As examples of suitable clocks, we will use two light sources which have a frequency ν_0 when they are at rest in an inertial frame in which there is no gravitational field. Consider the light emitted from Clock 1 at the time $t=0$, when the spaceship is instantaneously at rest in the inertial reference frame Σ at the time $t=0$. Let the light reach Clock 2 at a

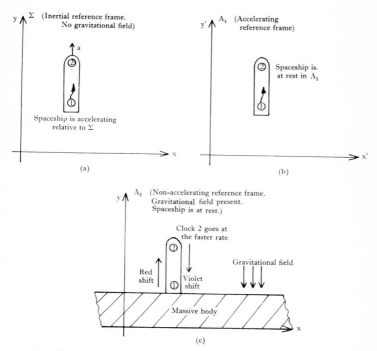

Figure 6.8. (a) In the inertial reference frame Σ the spaceship has an acceleration a. Light is emitted from two identical light sources 1 and 2 at the
time $t=0$, when the spaceship is instantaneously at rest in Σ. By the
time light from source 1 reaches source 2, source 2 is moving relative
to Σ, so that the light from source 1 will appear Doppler-shifted to an
astronaut by source 2. (b) In the accelerating reference frame A_1 in
which the spaceship is at rest, the light reaching source 2 from source 1
has a lower frequency than source 2. According to the principle of
equivalence, measurements in A_1 should be the same as in the non-
accelerating reference frame A_2 shown in (c), in which there is a gravitational field. Hence clock 2 in (c) should go at a faster rate than clock
1 in A_2 and visible light from light source 1 should be shifted to the
red end of the spectrum compared with the frequency of the exactly
similar light source 2.

time t later. By this time the Clock 2 has a speed $v=at$ and has travelled a distance $\frac{1}{2}at^2$ relative to Σ. Since light travels the distance $h+\frac{1}{2}at^2$, from Clock 1 to Clock 2, in the time t at speed c:

$$ct=h+\tfrac{1}{2}\,at^2.$$

125

If $\frac{1}{2}at^2 \ll ct$, then

$$t \simeq h/c; \quad v = at \simeq ah/c.$$

To an observer situated by Clock 2 it will appear as if the light reaching him from Clock 1 was emitted by a source moving away from him with a speed $v = ah/c$. He will observe the frequency of the light to be Doppler-shifted towards the red end of the spectrum. According to the non-relativistic theory of the Doppler effect (e.g. put $v \ll c$ in equation (4.55)), the frequency he observes is:

$$\nu' = \nu_0\left(1 - \frac{v}{c}\right) = \nu_0\left(1 - \frac{ah}{c^2}\right). \tag{6.2}$$

(Light emitted from Clock 2 at $t = 0$ appears to have been emitted by a source approaching Clock 1 with speed v, and the frequency of the light reaching Clock 1 from Clock 2 would appear to be $\nu_0(1 + ah/c^2)$.)

According to the principle of equivalence, these results should be the same as when the same experiments are carried out in the non-accelerating system A_2, shown in fig. 6.8 c in which the spaceship is at rest, and in which there is a gravitational field of such a strength that g, the acceleration due to the gravitational field, is numerically equal but opposite in direction to a, the acceleration of the spaceship relative to Σ. The light reaching Clock 2 from Clock 1 in fig. 6.8 c should have a frequency lower than Clock 2 by an amount $\nu_0 gh/c^2$, that is, should appear to be shifted to the red end of the spectrum, as shown in fig. 6.8 c. Conversely the frequency of light from Clock 2 reaching Clock 1 should be shifted to the violet end of the spectrum by $\nu_0 gh/c^2$. Now gh is the difference in gravitational potential between Clocks 1 and 2 in fig. 6.8 c. According to normal convention, Clock 2 is at a higher gravitational potential than Clock 1 in fig. 6.8 c, since work would have to be done to raise a mass from Clock 1 to Clock 2. In general, in equation (6.2), gh must be replaced by $\Delta\phi$ the difference in gravitational potentials between the clocks. Hence, if ν_1 and ν_2 are the frequencies of Clocks 1 and 2 respectively in fig. 6.8 c, from equation (6.2), if $\Delta\phi \ll c^2$:

$$\nu_2 \simeq \nu_1\left(1 + \frac{\Delta\phi}{c^2}\right) \tag{6.3}$$

and

$$\nu_1 \simeq \nu_2\left(1 - \frac{\Delta\phi}{c^2}\right). \tag{6.4}$$

Equations (6.3) and (6.4) should hold for other processes, such as the number of ticks of a clock per second. According to equation (6.3), in fig. 6.8 c Clock 2 should register more ticks than Clock 1.

126

If t_1 and t_2 are the time intervals measured by Clocks 1 and 2 respectively, since the time measured is proportional to the number of ticks, from equation (6.3):

$$t_2 = t_1 \left(1 + \frac{\Delta\phi}{c^2}\right).$$

Hence in fig. 6.8 c Clock 2 should go at a faster rate than Clock 1. All processes should take place at a faster rate near the top of the spaceship than near the bottom of the spaceship, which is at rest in the gravitational field in fig. 6.8 c.

For visible light, light coming from the higher altitude or higher gravitational potential is shifted to the violet end of the spectrum, and light going from the lower to higher altitude undergoes a red shift. This prediction was confirmed by Pound and Rebka in 1960, who measured the change in frequency of the $14 \cdot 4$ keV γ-ray line emitted by ^{57}Fe, over a vertical distance of 22 metre using the Mössbauer effect. Since $\Delta\phi = 9 \cdot 81 \times 22$,

$$\Delta\phi/c^2 = 9 \cdot 81 \times 22/(3 \times 10^8)^2 = 2 \cdot 4 \times 10^{-15}.$$

Hence the fractional shift in frequency should have been $2 \cdot 4 \times 10^{-15}$. Pound and Rebka's measured value was $0 \cdot 97 \pm 0 \cdot 04$ times the predicted value.

Another experimental check of equation (6.3) is to look at light coming from stars. This light comes from light sources at a lower gravitational potential than the surface of the earth. (Due to the larger mass of the star work would have to be done to take a mass from the star to the earth, even though over the last stage of the journey we would be losing some gravitational potential energy in the gravitational field of the earth.) From equation (6.3) it follows that the light from stars should be shifted to the red end of the spectrum. This is the gravitational red shift. Though the astronomical evidence is not conclusive, the results available are consistent with equation (6.3).

(c) Bending of the path of light in a gravitational field

Consider a spaceship moving with uniform acceleration a relative to an inertial reference frame Σ. If light enters a small hole in the spaceship, as shown in fig. 6.9 a, it travels in a straight line relative to Σ. However, relative to the accelerating spaceship, the light must appear to travel in a curved path, as shown in fig. 6.9 b. According to the principle of equivalence, observations in the accelerating spaceship should be the same as observations carried out by a stationary observer in a homogenous gravitational field, of such a strength that the acceleration due to gravity in the gravitational field is equal and opposite to the acceleration of the spaceship. Hence, light rays should be curved in a gravitational field, the light rays being bent in the direction of the acceleration due to gravity, as shown in fig. 6.9 c. As an

127

example, Einstein considered the deflection of light from a distant star, when the light passes close to the periphery of the sun. Measurements carried out during solar eclipses have confirmed Einstein's prediction. Near the earth the gravitational field is so weak that the deflection of light rays by the earth's gravitational field can be ignored.

It has been shown that in the presence of gravitational fields, light does not travel in straight lines, so that light rays cannot be used in these conditions to mark out a rectangular Cartesian co-ordinate system

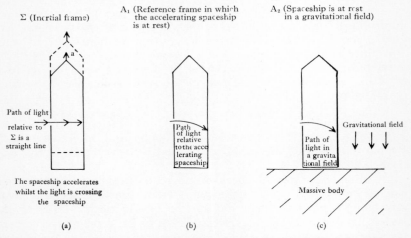

Figure 6.9. (*a*) Light enters an accelerating spaceship. The light path is a straight line relative to the inertial reference frame Σ. (*b*) Relative to the accelerating spaceship the path of the light is curved. (*c*) Hence, according to the principle of equivalence, light is deflected in a gravitational field.

in which Euclidean geometry can be applied. It has been shown that the rates of clocks depend on the gravitational potential. So do the measured lengths of rods. It is therefore not surprising to find that in the full general theory of relativity the numerical value of the speed of light depends on the gravitational potential. Hence the principle of the constancy of the speed of light must be modified in the presence of strong gravitational fields. One cannot use one single Cartesian reference frame with rulers and synchronized clocks to describe the *whole* universe. In the general theory of relativity, non-Euclidean geometry is used and the geometrical properties of space and time are allowed to vary in accordance with the distribution of matter. For terrestrial phenomena one can choose satisfactory approximations to inertial reference frames since the earth's gravitational field is very weak.

6.4 *Rotating reference frames and the principle of relativity*

Newton felt that absolute rotations could be observed. Newton's ideas can be illustrated in his own words describing his rotating bucket experiment, which is illustrated in fig. 6.10. In his *Principia Mathematica*, published in 1686, Newton wrote:

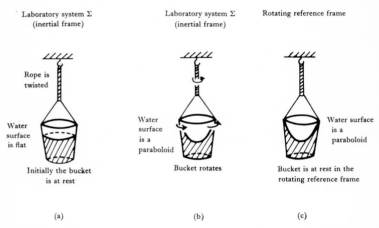

Laboratory system Σ
(inertial frame)

Laboratory system Σ
(inertial frame)

Rotating reference frame

Rope is twisted

Water surface is flat

Water surface is a paraboloid

Water surface is a paraboloid

Initially the bucket is at rest

Bucket rotates

Bucket is at rest in the rotating reference frame

(a)

(b)

(c)

Figure 6.10. (*a*) A bucket is suspended by a twisted rope. (*b*) When the bucket is released it rotates relative to the laboratory system Σ. Due to viscous forces the rotation of the bucket is communicated to the water. When the water is rotating, the surface of the water is a paraboloid. (*c*) In the rotating non-inertial reference frame in which the rotating bucket is at rest the surface of the water is still a paraboloid. In the rotating reference frame the shape of the ' stationary ' water is attributed to a centrifugal ' force '. On the basis of such effects in rotating reference frames Newton suggested that absolute rotations could be detected.

' The effects by which absolute and relative motions are distinguished from one another are centrifugal forces, or those forces in circular motion which produce a tendency of recession from the axis. For in a circular motion which is purely relative no such forces exist; but in a true and absolute circular motion they do exist, and are greater or less according to the quantity of the (absolute) motion.

' For instance. If a bucket, suspended by a long cord, is so often turned about that finally the cord is strongly twisted, then is filled with water, and held at rest together with the water; and afterwards by the action of a second force, it is suddenly set whirling about the contrary way, and continues, while the cord is untwisting itself, for some time in this motion, the surface of the water will at first be level, just as it was before the vessel began to move; but, subsequently, the vessel, by gradually communicating its motion to the water, will make it begin sensibly to rotate, and the water will recede little by little from the middle and rise up at the sides of the vessel, its surface assuming a concave form. (This experiment I have made myself.)

' ... At first, when the *relative* motion of the water in the vessel was *greatest*, that motion produced no tendency whatever of recession from the

5

axis; the water made no endeavour to move towards the circumference, by rising at the sides of the vessel, but remained level, and for that reason its *true* circular motion had not yet begun. But afterwards, when the relative motion of the water had decreased, the rising of the water at the sides of the vessel indicated an endeavour to recede from the axis; and this endeavour revealed the real circular motion of the water, continually increasing, till it had reached its greatest point, when *relatively* the water was at rest in the vessel . . . .

'It is indeed a matter of great difficulty to discover and effectually to distinguish the *true* from the apparent motions of particular bodies; for the parts of that immovable space in which bodies actually move, do not come under the observation of our senses.

'Yet the case is not altogether desperate; for there exist to guide us certain marks, abstracted partly from the apparent motions, which are the differences of the true motions, and partly from the forces that are the causes and effects of the true motions. If, for instance, two globes, kept at a fixed distance from one another by means of a cord that connects them, be revolved about their common centre of gravity, one might, from the simple tension of the cord, discover the tendency of the globes to recede from the axis of their motion, and on this basis the quantity of their circular motion might be computed. And if any equal forces should be simultaneously impressed on alternate faces of the globes to augment or diminish their circular motion, we might, from the increase or decrease of the tension of the cord, deduce the increment or decrement of their motion; and it might also be found thence on what faces forces would have to be impressed, in order that the motion of the globes should be most augmented; that is, their rear faces, or those which, in the circular motion, follow. But as soon as we knew which faces followed, and consequently which preceded, we should likewise know the direction of the motion. In this way we might find both the quantity and the direction of the circular motion, considered even in an immense vacuum, where there was nothing external or sensible with which the globes could be compared . . . .'

At first sight these arguments of Newton's seem extremely powerful. They were criticized, amongst others, by Ernst Mach. Mach's argument will be illustrated by a simple example. Consider a circular roundabout rotating inside a concentric circular room which has pictures painted on the walls. An observer on the rotating roundabout will tend to continue in a straight line relative to the fixed stars, and if he sits on the rotating roundabout he will feel himself tending to fly outwards. He will feel jerks on his feet as he tries to walk across the roundabout, and the observer would conclude that he was rotating. If the roundabout stopped rotating, and the circular wall started rotating in the opposite direction, when he looked at the wall the observer would see the same pictures on the circular wall rotating in the same directions as before, but he would no longer feel forces tending to push him outwards. He would conclude that the roundabout was rotating in the first instance, but not in the second. For example, he could attach a mass by a spring balance to the axis of the roundabout. If the mass were always at rest relative to the roundabout, the spring balance would register a force in the first instance but not in the second. However, the two cases considered so far are not really equivalent. In the first case not only is the roundabout rotating relative to the walls of the circular room, but it is rotating relative to all the rest of the matter in the universe. To be completely equivalent all the masses in the universe would have to rotate around the stationary roundabout in the second case. According to *Mach's Principle* if they did so, due to the changes in gravitational forces due to the motions of the matter in the universe, in the second case the observer on the stationary roundabout would experience gravitational forces pulling him outwards. The centrifugal and Coriolis ' forces ' are interpreted in the rotating reference frame as gravitational forces associated with the rotation of the masses of the universe. If this were correct the observer on the rotating roundabout would not be able to say *experimentally* whether the roundabout was rotating relative to the rest of the matter of the universe or whether the roundabout was at rest, and *all* the rest of the matter in the universe was rotating relative to a stationary roundabout. In this way we can develop a general theory of relativity in which gravitational forces play an important part. On this view one cannot say by *experiment* whether Ptolemy's view, that the earth is at rest and the sun and all the stars rotate about the earth, or Copernicus' view, that the earth rotates about its axis and around the sun, is correct. Common sense tells us that the latter view is the more likely one, but how can we find out experimentally? In reference frames undergoing linear accelerations the inertial forces can also be interpreted as gravitational forces arising from the accelerations of the rest of the matter in the universe relative to the accelerating reference frame.

By analogy with the electromagnetic case, where the electric and magnetic forces between moving charges depend on the velocities and accelerations of the charges, it is not unreasonable to assume that the gravitational forces between moving masses *may* depend on their velocities and accelerations. However, due to their great distance from the earth, it might appear at first sight impossible for distant stars to affect terrestrial phenomena. As an analogy we shall consider *Olbers' Paradox*. The intensity of light reaching the earth from a stationary star should vary as $1/r^2$, where r is the distance of the star from the earth. If it is assumed that the stars are distributed uniformly in the universe, the number of stars at a distance between r and $(r + dr)$ from the earth is the number in a spherical shell of volume $4\pi r^2 dr$ so that, if the intensity of light from a star is proportional to $1/r^2$, the total light from stars at a distance between r and $(r + dr)$ from the earth is proportional to $4\pi r^2 dr \times (1/r^2)$, that is, proportional to dr. If one integrates to $r = \infty$, the total amount of light reaching the earth from the stars in the universe should be infinite. Why are we not blinded every time we look at the sky? Olbers' paradox can be resolved by the fact that the distant stars in the universe are going away from the earth at high speeds (the expanding universe). When a light source is moving, due to the Doppler effect, discussed in § 4.7, the frequency of the light, and hence the energy of the photons and the intensity of the light emitted in the direction opposite to the direction of motion of the light source, is reduced. Due to the expansion of the universe, the intensity of light reaching the earth from the distant stars is therefore reduced, since the stars are moving away from the earth. Hence, the total intensity of the light from stars reaching the earth is finite. Similarly, if the distant stars gave gravitational effects, whose intensity varied with distance r from the star as $1/r^n$, with $n \leqslant 2$, the combined effect of all the stars in the universe could be infinite. The actual total gravitational effects due to the stars in the universe depends on the boundary conditions, such as the size of the universe, the recession of distant nebulae, etc. All we have tried to show is that distant matter in the universe might conceivably have an effect on terrestrial phenomena. Some people go so far as to say that the inertial masses of bodies arise in this way.

In this section we have indicated only how one can move towards a principle of relativity for all reference frames, whether they are accelerating or not, and we have shown the importance of gravitational forces in such a scheme. A reader interested in more comprehensive popular accounts is referred to:

> *Introductory Relativity* by W. G. V. Rosser (Butterworths, London, 1967);
> *Readable Relativity* by C. V. Durrell (Bell, London, 1931);
> *Space, Time and Gravitation* by A. S. Eddington (Cambridge University Press, 1920);
> *The Unity of the Universe* by D. W. Sciama (Faber and Faber, London, 1959).

Problems

6.1. What would the length of the day have to change to if the apparent acceleration due to gravity at the Equator is to be zero? Take the radius of the earth to be $6 \cdot 37 \times 10^6$ m. (Hint: Put $g = 0$ in equation (6.1) to calculate ω. Assume $g_0 = 9 \cdot 81$ m s^{-2}).

6.2. Calculate the fractional difference in frequency between the frequency of an atomic process on the top of the spire of Salisbury Cathedral, altitude 124 m, and the frequency of the same process at ground level. Take g to be $9 \cdot 81$ m s^{-2}. (Hint: Use equation (6.3).)

6.3. Calculate the difference in the readings of clocks at sea level and at the top of Mount Everest respectively after one year. Take $g = 9 \cdot 81$ m s^{-2} and the altitude of Mount Everest as 8842 m. (Hint: Use equation (6.3).)

CHAPTER 7
the clock paradox

7.1 *Space travel*

So many of the stars in the universe are many millions of light years away that, at first sight, it would seem impossible to reach very many of them in a normal life span. However, this ignores the phenomena of time dilatation and length contraction. As an example we shall consider a star *at rest* relative to the earth at a distance of 1000 light years from the earth. At first sight it might appear that it would take a spaceship travelling at the speed of light 1000 years to get there. This however is the time measured in the laboratory system. If k is the number of seconds in a year, and c is the speed of light, the distance of the star from the earth is 10^3 kc metre, measured in the laboratory system Σ, in which both the earth and the star are at rest, as shown in fig. 7.1 a. Let a spaceship move away from the earth, in a direction opposite to the star, let the spaceship turn around and then accelerate until it reaches a speed v relative to the earth. Let the spaceship then continue with uniform velocity v relative to the earth (Σ). Let the spaceship pass the earth (Event 1), when the spaceship is moving with uniform velocity v at a

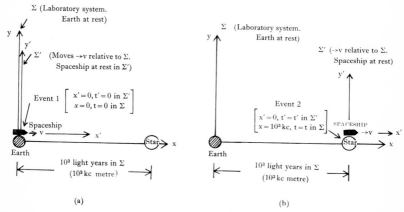

(a)   (b)

Figure 7.1. (*a*) The earth is at the origin of Σ and the spaceship is at the origin of Σ'. The spaceship leaves the earth at Event 1 at $x = x' = 0$ at $t = t' = 0$ and travels with uniform velocity v relative to the earth to go to the star 10^3 light years away. (*b*) The spaceship reaches the star at Event 2 at $x = 10^3 kc$ metre at a time t in Σ. Event 2 is at $x' = 0$ at a time t' in Σ'.

133

time $t=0$ on a clock on the earth and a time $t'=0$ on a clock on the spaceship. Let the earth be at the origin of the inertial reference frame Σ, and let the spaceship be at rest at the origin of the inertial reference frame Σ', when the spaceship is moving away from the earth towards the star with uniform velocity v, as shown in fig. $7.1\,a$. The co-ordinates of Event 1 in Σ are $x=0$, $t=0$, and in Σ' $x'=0$, $t'=0$, as shown in fig. $7.1\,a$.

Let the spaceship reach the star (Event 2) at a time t measured relative to the earth (Σ) and a time t' measured by the clock on the spaceship (Σ'). Relative to the laboratory system Σ in which the earth is at rest, the co-ordinates of Event 2 (the spaceship's reaching the star) are:

$$x = 10^3\,kc; \qquad t=t,$$

and relative to the spaceship (Σ'):

$$x'=0; \qquad t'=t',$$

as shown in fig. $7.1\,b$. From the Lorentz transformations:

$$t = \gamma(t' + vx'/c^2).$$

For Event 2, $x'=0$. Hence:

$$t = \gamma t' = t'/\sqrt{(1-v^2/c^2)}. \qquad (7.1)$$

Since $\gamma > 1$, $t' < t$. The time t' of the journey measured by the clock on the spaceship is less by a factor γ than the time t for the journey measured relative to the earth. Equation (7.1) is the normal expression for time dilatation. The time t' between Events 1 and 2 is the proper time interval between the event of the spaceship's leaving the earth and the event of the spaceship's reaching the star, measured by one clock, the clock on the spaceship. The time t for the journey relative to the earth (Σ) must be measured by the radar methods described in § 4.2, or by using two synchronized spatially separated clocks, one on the earth and one on the star.

As a numerical example it will be assumed that the spaceship reaches the star in a time $t'=10$ years $=10k$ second, measured by the clock on the spaceship. If v is the speed of the spaceship relative to the earth, the spaceship goes with a speed v for a time $t=\gamma t'$ relative to the earth (Σ) covering a distance $v\gamma t'$, relative to the earth (Σ), before reaching the star. Since the star is $10^3 kc$ metre from the earth, measured relative to the laboratory system (Σ), relative to Σ we have:

$$10^3 kc = v\gamma t'. \qquad (7.2)$$

If $t'=10k$ second:

$$10^3 kc = v\gamma 10k$$

134

or

$$\frac{v}{c} = \frac{10^3}{10\gamma} = \frac{100}{\gamma} = 10^2 \left(1 - \frac{v^2}{c^2}\right)^{1/2}.$$ (7.3)

Squaring:

$$\frac{v^2}{c^2} = 10^4 - 10^4 \frac{v^2}{c^2},$$

$$(1 + 10^4)v^2/c^2 = 10^4,$$

$$\frac{v^2}{c^2} = \frac{10^4}{(1 + 10^4)} = \frac{1}{(1 + 10^{-4})} = (1 + 10^{-4})^{-1}.$$

Using the binomial theorem:

$$\frac{v}{c} = (1 + 10^{-4})^{-1/2} = 1 - 0.5 \times 10^{-4},$$

$$v = 0.99995c.$$

Substituting for v in equation (7.3):

$$\gamma = \frac{100}{0.99995} = 100.005.$$ (7.4)

The above analysis shows that, if the spaceship goes at a speed of $0.99995c$ relative to the earth, it will reach the star in 10 years measured by the clock on the spaceship. Relative to the earth the spaceship takes a time $t = \gamma t' = 1000.05$ years travelling at a speed $0.99995c$ to reach the star. Due to the fact that the proper time interval t' measured by a clock on the spaceship is less by a factor γ than the improper time interval t measured in the laboratory system, it is possible to reach the star 1000 light years away from the earth (measured relative to the laboratory system Σ) in a time of 10 years, measured by the clock on the spaceship (Σ').

Relative to the spaceship, which is at rest at the origin of Σ', the earth passes the spaceship at Event 1 at $x' = 0$, $t' = 0$. The earth and star then move with a velocity $v = 0.99995c$ along the *negative* x' axis of Σ' for 10 years or $10k$ second, the star reaching the spaceship at Event 2 at $x' = 0$, $t' = t'$ relative to Σ'. The astronaut on the spaceship calculates the distance between the earth and the star to be $l' = vt'$, relative to Σ'. On the other hand, relative to the earth (Σ), the spaceship goes at a speed v for a time $t = \gamma t'$ covering a distance $v\gamma t'$ before reaching the star. Hence the *proper* distance between the earth and the star, that is, the distance between them measured in the inertial reference frame Σ in which they are both at rest is $l_0 = \gamma vt'$. Notice $l = l_0/\gamma$. Thus, relative to the moving spaceship, the distance between the earth and the star is Lorentz contracted in the direction of motion of the earth relative to the

135

spaceship (Σ'). This example illustrates how time dilatation and length contraction are intimately connected.

A proton of rest mass 938 MeV/c^2 of velocity $0\cdot99995c$ such that $\gamma = 100\cdot005$ has an energy of $\gamma m_0 c^2 \simeq 100 \times 938$ MeV $= 9\cdot38 \times 10^{10}$ eV. There are plenty of cosmic ray protons of this energy moving about in interstellar space. Actually protons of energy up to 10^{19} eV have been observed in the cosmic radiation. Provided the deflection of the protons by galactic magnetic fields can be ignored, the analysis of this section is applicable to cosmic ray protons. Another numerical example is given in Problem 7.1.

Now the interesting question is, if, after the time $t' = 10$ years on the spaceship, the spaceship turned around quickly and came back to the earth with the same uniform velocity v, would the time for the complete journey be 20 years, as measured by the clock on the spaceship and $2 \times 1000\cdot05 = 2000\cdot1$ years relative to the clock on the earth? According to Einstein[1] the above suggestion should be correct, and the astronaut should come back to the earth 20 years older than when he left, whereas his contemporaries on the earth would have died many centuries earlier. This suggestion is known as the clock paradox. It will be shown in § 7.2 that the clock paradox is consistent with the postulates of special relativity, and in § 7.3 that it is consistent with experiments. However, before we rush to look for spaceships in the search for eternal youth, we shall consider the fuel requirements to drive such a spaceship. (Incidentally, everything else on the spaceship would go at the slower rate given by equation (7.1), so that the astronaut would not notice he was living longer, until he returned to the earth.)

Let a rocket of initial rest mass m_i emit gases at a speed w relative to the rocket until the rocket reaches a speed u, when its final rest mass is m_f. It can be shown (Rosser[2 a]) that

$$\frac{m_i}{m_f} = \left(\frac{c+u}{c-u}\right)^{c/2w}.$$

With present-day rockets w, the exhaust speed of the gases, is of the order of 10 kilometre per second, so that $c/2w$ is $3 \times 10^8/2 \times 10^4 = 1\cdot5 \times 10^4$. If the final speed of the rocket were, say, $u = c/2$, then

$$\left(\frac{m_i}{m_f}\right) = \left(\frac{c+c/2}{c-c/2}\right)^{1\cdot5 \times 10^4} = (3)^{1\cdot5 \times 10^4} \simeq 10^{7157}$$

If the final mass of the rocket is to be large enough to hold an astronaut, say one ton, one would have to start with a spaceship weighing 10^{7157} tons. It can be seen that with present-day rocketry techniques it would be impossible to accelerate a rocket to a speed comparable to the speed of light. If, however, we could use photons (light quanta) or high speed atomic particles as exhaust 'gases', then $w \simeq c$, and for $u = c/2$, m_i/m_f would only be $\sqrt{3}$. Hence there is room for considerable improvement.

At present, the only particles we can accelerate to very high speeds are atomic particles. For example, a proton of energy 30 GeV from the CERN proton synchrotron has a speed of $0.9995c$. Hence the predictions of the clock paradox are important for atomic particles, accelerated in the laboratory, and an experimental check of the clock paradox using atomic particles in the laboratory will be described in § 7.3.

7.2 Imaginary experiment on the clock paradox

Let a spaceship leave the earth at a time $t = 0$ on a clock on the earth and a time $t' = 0$ on a clock on the spaceship. Let the spaceship travel with uniform velocity v relative to the earth (laboratory system Σ). Let the spaceship turn around at a time $t_L/2$ relative to the laboratory system (Σ), when it has covered a distance $vt_L/2$ relative to the earth (Σ), as illustrated in the displacement time curve in fig. 7.2 for the special case when $v = 0.6c$. Let the spaceship turn around quickly, and let it return with uniform velocity v relative to the earth and let it reach the earth at a time t_L, measured by the clock on the earth, as shown in figure 7.2. Let the time of the complete journey be measured to be t_S' by a clock on the spaceship, so that the spaceship turns around at a time $t_S'/2$ on the clock on the spaceship. Let radio (or light) signals be sent out from the spaceship at time intervals of T_0', measured by the clock on the spaceship as shown in fig. 7.2. The frequency of these signals measured by the clock on the spaceship is $\nu_0' = 1/T_0'$. The total number of signals transmitted from the spaceship before turning around at the time $t_S'/2$, relative to the spaceship, is $(t_S'/2) \div T_0' = t_S'/2T_0' = \nu_0' t_S'/2$. When the spaceship turns around at a time $t_L/2$ relative to the laboratory system Σ, it is at a distance $vt_L/2$ from the earth, measured relative to the laboratory system Σ, as shown in fig. 7.2. The radio signal emitted from the spaceship at the time $t_L/2$, relative to the laboratory system, covers the distance $vt_L/2$ to the earth at a speed c in a time $vt_L/2c$ reaching the earth at a time $t_L/2 + vt_L/2c$ or $\frac{1}{2}t_L(1 + v/c)$ on the earth clock, as illustrated in fig. 7.2. According to equation (4.9), when the spaceship is moving away from the earth, signals emitted at time intervals of T_0' from the spaceship (measured by the spaceship clock) reach the earth at time intervals of $KT_0' = K/\nu_0'$ (measured by the earth clock). Since the total number of signals emitted from the spaceship before it turns around is $\nu_0' t_S'/2$, and the time interval between each signal's reaching the earth is K/ν_0' on the earth clock, the signal emitted from the spaceship, when it turns around, must reach the earth at a time $\frac{1}{2}\nu_0' t_S' \times K/\nu_0'$ on the earth clock. Hence:

$$\frac{t_L}{2}\left(1 + \frac{v}{c}\right) = \frac{\nu_0' t_S'}{2} \times \frac{K}{\nu_0'} = \frac{t_S' K}{2}. \qquad (7.5)$$

The $\nu_0' t_S'/2$ signals, emitted from the spaceship between the time the spaceship turns around and the time the spaceship returns to the earth,

137

reach the earth between the time $\frac{1}{2}t_L(1+v/c)$ and t_L on the earth clock, that is, in a time interval of $\frac{1}{2}t_L(1-v/c)$ on the earth clock, as shown in fig. 7.2. It was shown in § 4.7 that, if the spaceship emits signals of frequency ν_0', and, if the frequency of the signals reaching the earth is ν_0'/K, when the spaceship is going away from the earth with uniform velocity v, then the frequency of the signals reaching the earth when the spaceship is moving towards the earth with uniform velocity v is $K\nu_0'$.

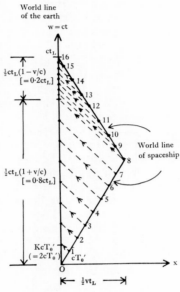

Figure 7.2. Imaginary experiment on the clock paradox. The speed of the spaceship in this figure is $0 \cdot 6c$ relative to the earth, so that in this special case $\gamma = 1/\sqrt{(1-v^2/c^2)} = \frac{5}{4}$ and $K = \sqrt{(1+v/c)}/\sqrt{(1-v/c)} = 2$. From equation (7.7) $t_L = \gamma t_S' = \frac{5}{4}t_S'$. The spaceship transmits 16 radio signals at time intervals of T_0' relative to the spaceship. The first eight signals reach the earth at time intervals of $KT_0' = 2T_0'$ on the earth clock, in a total time of $16T_0'$ on the earth clock. The second eight signals reach the earth at time intervals of $T_0'/K = T_0'/2$ on the earth clock in a total time interval of $8 \times T_0'/2 = 4T_0'$ on the earth clocks. Thus the total time of the journey is $20T_0'$ measured by the earth clock and $16T_0'$ measured by the spaceship clock. This is in agreement with equation (7.7).

Hence the time between each successive signal's reaching the earth is $1/K\nu_0'$, for signals emitted from the spaceship when it is moving towards the earth. Since the total number of signals emitted from the spaceship between its turning around and returning to the earth is $\frac{1}{2}\nu_0't_S'$ and the time interval between each signal's reaching the earth is $1/K\nu_0'$ on the earth clock, the total time for which these signals are received on the

138

earth is $\frac{1}{2}v_0't_S' \times (1/Kv_0')$ measured by the earth clock. Hence:

$$\frac{t_L}{2}\left(1-\frac{v}{c}\right) = \frac{v_0't_S'}{2} \times \frac{1}{Kv_0'} = \frac{t_S'}{2K}. \tag{7.6}$$

Multiplying equations (7.5) and (7.6) gives:

$$t_S'^2 = t_L^2(1 - v^2/c^2)$$

or

$$t_L = \frac{t_S'}{\sqrt{(1 - v^2/c^2)}} = \gamma t_S'. \tag{7.7}$$

Equation (7.7) shows that the time for the journey, measured by the clock on the spaceship, should be less than the time for the journey measured by a clock at rest on the earth. The above example illustrates clearly the possibility that a person who goes on a journey into outer space and back at a speed comparable to c will age less than a person who stays on the earth. In the example discussed in § 7.1 we have $t_S' = 20$ years for the journey to the star and back. If $v = 0{\cdot}99995c$, such that $\gamma = 100{\cdot}005$, then t_L is $2000{\cdot}1$ years.

Dividing equation (7.5) by equation (7.6) gives:

$$\frac{1+v/c}{1-v/c} = K^2; \quad \text{or} \quad K = \sqrt{\left(\frac{1+v/c}{1-v/c}\right)}.$$

This is in agreement with equation (4.16).

For a discussion of the case when radio signals are transmitted from the earth and received by the spaceship the reader is referred to Rosser[2b].

7.3 *Experimental check of the clock paradox**

If one started with N_0 radioactive atoms at rest on the earth and N_0 identical atoms at rest in a spaceship, when the spaceship goes on a journey of the type illustrated in fig. 7.2 then, when the spaceship returns to the earth, according to the law of radioactive decay, the number of radioactive atoms left in the laboratory system (denoted N_{LAB}) should be given by:

$$N_{LAB} = N_0 \exp\left(-t_L/T_0\right), \tag{7.8}$$

where t_L is the time for the journey measured by the earth clock, and the number of radioactive atoms left in the spaceship at the end of the journey should be:

$$N_{SPACESHIP} = N_0 \exp\left(-t_S'/T_0\right), \tag{7.9}$$

where t_S' is the time of the journey measured by the clock on the spaceship, and T_0 is the mean life of the radioactive atoms when they are at rest. If $t_S' < t_L$, as given by equation (7.7), then $N_{SPACESHIP}$ should be greater than N_{LAB} and there should be more radioactive atoms left undecayed in the spaceship than on the earth. Using equation (7.7), equation (7.9) becomes:

$$N_{\text{SPACESHIP}} = N_0 \exp\left(-t_S'/T_0\right) = N_0 \exp\left(-t_L/\gamma T_0\right). \qquad (7.10)$$

An experiment of the above type was carried out by Farley and collaborators[3] using a μ-meson storage ring at CERN. Many experiments on μ-mesons have shown that the mean life of μ-mesons is

$$T_0 = 2 \cdot 20 \times 10^{-6} \text{ second}$$

when they are at rest, so that for μ-mesons at rest in the laboratory, equation (7.8) becomes:

$$N_{\text{LAB}} = N_0 \exp\left(-t_L/2 \cdot 20 \times 10^{-6}\right). \qquad (7.11)$$

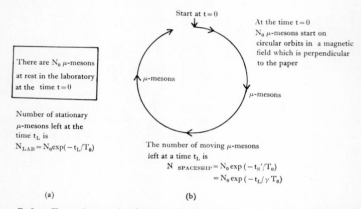

LABORATORY SYSTEM Σ

Start at $t = 0$

At the time $t = 0$
N_0 μ-mesons start on circular orbits in a magnetic field which is perpendicular to the paper

There are N_0 μ-mesons at rest in the laboratory at the time $t = 0$

μ-mesons

μ-mesons

Number of stationary μ-mesons left at the time t_L is
$N_{\text{LAB}} = N_0 \exp(-t_L/T_0)$

The number of moving μ-mesons left at a time t_L is
$N_{\text{SPACESHIP}} = N_0 \exp(-t_S'/T_0)$
$= N_0 \exp(-t_L/\gamma T_0)$

(a)

(b)

Figure 7.3. Experimental check of the clock paradox using μ-mesons. According to equation (7.10), at a time t_L in the laboratory system, $N_{\text{SPACESHIP}}$ should be greater than N_{LAB}. This prediction is confirmed by experiment.

Instead of putting the μ-mesons in a spaceship, in the experiment at CERN, Farley *et al.* sent the μ-mesons around in circular orbits in a magnetic field in the laboratory, as illustrated in fig. 7.3 b. They found that the rate of decay of μ-mesons of momenta $1 \cdot 274$ GeV$/c$ going around in circles in a magnetic field satisfied the experimental relation:

$$N_{\text{SPACESHIP}} = N_0 \exp\left[-t_L/(26 \cdot 15 \pm 0 \cdot 03) \times 10^{-6}\right], \qquad (7.12)$$

where t_L is the time of the μ-meson journey relative to the laboratory. It can be seen that μ-mesons which go on a journey 'into outer space' (that is, in this case, move in a circle of 5 metre diameter) live longer on

140

the average than ' stay at home ' μ-mesons at rest in the laboratory. This confirms the clock paradox. In this case the μ-mesons are accelerated continuously as they move in circles. However, by treating the circle as a series of straight line sections, equation (7.7) should be applicable to μ-mesons going in circles. The magnitude of the momenta of the μ-mesons was $1 \cdot 274 \text{ GeV}/c$ corresponding to $\gamma = 12 \cdot 14$, so that $\gamma T_0 = 12 \cdot 14 \times 2 \cdot 20 \times 10^{-6} = 26 \cdot 72 \times 10^{-6}$ second. Hence, according to equation (7.10), one would have expected:

$$N_{\text{SPACESHIP}} = N_0 \exp{(-t_{\text{L}}/26 \cdot 72 \times 10^{-6})}. \qquad (7.13)$$

The mean life of $26 \cdot 15 \times 10^{-6}$ in equation (7.12) is 2% less than the predicted value of $26 \cdot 72 \times 10^{-6}$ in equation (7.13). This 2% was interpreted as a loss of μ-mesons due to imperfections in the magnetic field. For a discussion of other experimental evidence in favour of the clock paradox the reader is referred to Rosser[2 b]. All the evidence available at present is consistent with equation (7.7).

7.4 Discussion

So far in this chapter we have only considered the effects of motion relative to an inertial reference frame. It was illustrated in § 6.3 that changes in gravitational potential affect the rate of a clock. By comparison with fig. 6.8 c it can be seen that the higher up a clock is in the earth's gravitational field, the faster the rate of the clock should be, so that, due to the increase in gravitational potential, an astronaut in a satellite should age more than somebody at sea level, whereas due to his motion relative to the earth, according to equation (7.7), the astronaut should age less. It can be shown (Rosser[2 c]) that, if the altitude of a satellite in a circular orbit around the earth is $> R/2$ ($> 3200 \text{ km}$ above sea level), where R is the radius of the earth, then the effect of the change in gravitational potential predominates, and the astronaut should age more than a sea level person, whereas if the altitude of the satellite is $< R/2$, the effect associated with equation (7.7) predominates and the astronaut should age less than a person at sea level. The speeds of satellites are ~ 10 kilometre per second, which is only $\sim 3 \cdot 3 \times 10^{-5}c$. Using equation (7.7), the reader can check that the difference $(t_{\text{L}} - t_{\text{S}}')$ is only $\sim 5 \times 10^{-5}$ second per day for such a satellite. To obtain large effects the satellite (or spaceship) would have to go at speeds comparable to the speed of light.

If the fuel problems discussed in § 7.1 could be overcome, the most comfortable way of going on a space journey would be to control the rate of the exhaust of gases, such that the spaceship always went at an acceleration $g = 9 \cdot 81$ metre per second² relative to the inertial reference frame in which the spaceship is instantaneously at rest. It follows from the discussion of § 6.3 that the conditions in the spaceship would then be similar to conditions on the earth. At such an acceleration, in one year (relative to the earth), the spaceship would travel about 0.43

141

light years and reach a speed of $0.75c$ relative to the earth. If then the spaceship decelerated for one year (relative to the earth) with deceleration $g = 9.81$ metre per second^2 measured relative to the inertial reference frame in which the spaceship is instantaneously at rest, relative to the earth the spaceship would reach a point 0.86 light years away from the earth in two years, relative to the earth. If it came back in the same way the spaceship could do the journey there and back in four years (measured by a clock on the earth) and in a time of three-and-a-half years according to the clock on the spaceship.

References

(1) EINSTEIN, A., *Annalen der Physik.*, **17**, (1905), 891.
(2) ROSSER, W. G. V., *Introductory Relativity* (Butterworths, London, 1967). (a) p. 174;   (b) Chap. 8;   (c) p. 272.
(3) FARLEY, F. J. M., BAILEY, J., PICASSO, F., *Nature, Lond.*, **217**, (1968), 17.

Problems

7.1. The diameter of our galaxy is about 10^5 light year. How long does it take a proton (in the proton rest frame) to cross the galaxy if its energy is 10^{19} eV? Take the rest mass of the proton to be 938 MeV/c^2. Ignore the effects of galactic magnetic fields, which in practice would deflect such a proton.

7.2. A spaceship accelerates quickly and then moves with uniform velocity relative to the solar system, until it reaches a star 8 light year away. A clock on the spaceship records the time for the journey as six year. Calculate (a) the speed of the spaceship relative to the earth and (b) the time of the journey relative to the earth. (c) If the spaceship turns around quickly and returns to the earth with the same velocity as on the outward journey, compare the total time for the journey relative to the earth and relative to the spaceship.

7.3. On her 29th birthday a lady physicist concludes that she would like to remain 29 for at least 10 year. She decides to go on a journey into outer space with uniform velocity. What is the minimum speed she must move relative to the laboratory so that she can return 10 year later (relative to the laboratory) and still say, quite truthfully, that she is only 29?

7.4. A total of 2000 charged π-mesons are created at the origin of an inertial frame. Half of the π-mesons remain at rest at the origin, whilst the other half go on a journey with a uniform speed of $0.995c$. After a path length of 15 m, the travelling π-mesons are deflected in a magnetic field so that they return back to the origin.

(a) How many of the moving π-mesons should survive the journey?

(b) How many of the stationary π-mesons are left at the origin when the travelling π-mesons return?

Take the mean lifetime of charged π-mesons to be equal to 2.5×10^{-8} s when they are at rest.

142

ANSWERS TO PROBLEMS

1.2. (a) 25·5 m s^{-1} in a direction tan^{-1} 5 East of North.
(b) 27·8 m s^{-1} in a direction tan^{-1} 6·3 East of North.
1.3. (a) $x' = 0$; $y' = -19\cdot6$ m; (b) $x = 30$ m; $y = -19\cdot6$ m.
2.1. m/m_0 is (a) $1 + 3\cdot9 \times 10^{-14}$; (b) $1 + 3\cdot9 \times 10^{-12}$; (c) 1·005;
(d) 1·155; (e) 2·294; (f) 7·089; (g) 22·4; (h) 71.
2.2. v/c is (a) 0·196; (b) 0·549; (c) 0·94.
m/m_0 is (a) 1·02; (b) 1·197; (c) 2·97.
2.3. v/c is (a) 0·145; (b) 0·42; (c) 0·87.
2.4. (a) 0·536 mm; (b) 2·27 mm; (c) 11·95 mm; (d) 38·06 mm.
2.5. (a) 0·079 MeV; (b) 0·261 MeV; m/m_0 is (a) 1·15; (b) 1·51.
2.6. (a) $2\cdot56 \times 10^5$ V; (b) 0·75c.
2.7. 144 MeV/c^2.
2.8. 930 MeV/c^2.
4.1. (a) (i) $x' = (5 - 1\cdot35 \times 10^9)$m; $t' = (7\cdot5 - 10^{-8})$ s.
(ii) $x' = 4\cdot25 \times 10^8$ m; $t' = 0\cdot75$ s.
(iii) $x' = 7\cdot43 \times 10^{10}$ m; $t' = -146$ s.
(b) (i) $x = (12\cdot5 + 9 \times 10^8)$ m; $t = (5 + 2\cdot5 \times 10^{-8})$ s.
(ii) $x = 1\cdot21 \times 10^{10}$ m; $t = 27\cdot5$ s.
(iii) $x = 1\cdot36 \times 10^{11}$ m; $t = 312\cdot5$ s.
4.2. $6\cdot37 \times 10^{-2}$ m.
4.3. 0·87c.
4.4. (a) 1·36 km; (b) 4·6 km; (c) 14·5 km.
4.5. (a) 0·87c; (b) 300 kg.
4.6. (a) 453 nm; (b) 115 nm.
4.7. £$1\cdot85 \times 10^8$.
4.8. (a) 0·5c; (b) 5·2 s.
4.9. (a) $x' = 0\cdot75c$; $t' = 2\cdot75$ s; $v = 0\cdot75c$.
(b) $x' = 2\cdot75c$; $t' = 0\cdot75$ s; $v = 0\cdot75c$.
5.1. (a) $u_x = 30(1 - 2\cdot22 \times 10^{-15})$m s^{-1}; $u_y = 0$.
(b) $u_x = 20$ m s^{-1}; $u_y = 10(1 - 2\cdot22 \times 10^{-15})$ m s^{-1}.
5.2. (a) $u_x = 0\cdot917c$; $u_y = 0$; (b) $u_x = 0\cdot1c$, $u_y = 0\cdot896c$.
5.3. $u_x'' = 0\cdot87c$; $u_y'' = 0\cdot33c$.
6.1. 1·4 hour.
6.2. $\Delta\nu/\nu = 1\cdot35 \times 10^{-14}$.
6.3. 3×10^{-5} s.
7.1. 296 s.
7.2. (a) 0·8c; (b) 10 year; (c) $t_{LAB} = 20$ year, $t_{SPACESHIP} = 12$ year.
7.3. 0·995c.
7·4. (a) 670; (b) 18.

INDEX

THE WYKEHAM SCIENCE SERIES
for schools and universities

1 *Elementary Science of Metals* J. W. MARTIN and R. A. HULL
   (S.B. No. ...)

2 *Neutron...* G. E. Thompson... G. R. NOAKES
   (S.B. No. 85109 0..6)

3 *Essentials of Meteorology* D. H. McINTOSH,
   (S.B. No. 85109 04. 0) A. S. THOM and V. T. SAUNDERS

4 *Nuclear Fusion* H. R. HULME and A. McB. COLLIEU
   (S.B. No. 85109 0..8)

5 *Water Waves* N. F. Barber and G. GHEY
   (S.B. No. 85109 0.. .)

6 *Gravity and the Earth* A. H. Cook and V. T. SAUNDERS
   (S.B. No. 85109 0..0)

7 *Relativity and High Energy Physics* W. G. V. ROSSER
   (S.B. No. 85109 08. X) and R. E. McCULLOCH

Price per book for the Science Series 20s.—£1.00 net *in U.K. only*

THE WYKEHAM TECHNOLOGICAL SERIES
for universities and institutes of technology

1 *Frequency Conversion* J. THOMSON,
   (S.B. No. 85109 030 3) W. E. TURK and M. BEESLEY

Price per book for the Technological Series **25s.**—**£1.25 net** *in U.K. only*

*Standard Book Catalogue numbers.

148